Buckets from an English Sea

Buckets from an English Sea

1832 and the Making of
Charles Darwin

Louis B. Rosenblatt

OXFORD
UNIVERSITY PRESS

OXFORD
UNIVERSITY PRESS

Oxford University Press is a department of the University of Oxford. It furthers
the University's objective of excellence in research, scholarship, and education
by publishing worldwide. Oxford is a registered trade mark of Oxford University
Press in the UK and certain other countries.

Published in the United States of America by Oxford University Press
198 Madison Avenue, New York, NY 10016, United States of America.

© Oxford University Press 2018

Library of Congress Cataloging-in-Publication Data
Names: Rosenblatt, Louis Barry, author.
Title: Buckets from an English Sea : 1832 and the making
of Charles Darwin / Louis B. Rosenblatt.
Description: New York, NY : Oxford University Press, [2018] |
Includes bibliographical references and index.
Identifiers: LCCN 2017030144 | ISBN 9780190654405
Subjects: LCSH: Darwin, Charles, 1809–1882. | Beagle Expedition (1831–1836) |
Naturalists—England—Biography. | Evolution (Biology)—History.
Classification: LCC QH31.D2 R57 2018 | DDC 576.8/2092 [B]—dc23
LC record available at https://lccn.loc.gov/2017030144

1 3 5 7 9 8 6 4 2
Printed by Sheridan Books, Inc., United States of America

CONTENTS

LIST OF ILLUSTRATIONS

ACKNOWLEDGMENTS

I am indebted to so many. To all my former students who sought more than I could readily offer, so I had to push what I knew. To former teachers, like Drs. Hannaway, Pocock, and the Fabers, who engaged what I offered. And to a myriad of others whom I only knew through their writings, like Popper and Havelock, who offered clarity of judgment where I had only seen "stuff." And thanks to those who read early drafts: Bobby Loftus, Roy Baxandall, Dave Kinne, Marshall Gordon, and especially Dale Beran. I also want to thank my family, grown now beyond the four of us to include Shanna, Gabe, and Nora, for all our talks and walks. And finally Bonnie, companion, critic, and lovely at both.

Buckets from an English Sea

1

PRELUDE: A WORD ON
BEGINNINGS AND ENDS

History is mostly clouds and clouds don't have doors. But every once in a while transition becomes visible change, a sense that we are passing through a doorway. The year 1832 was such a moment in England and in the life of the young Charles Darwin. Hence the two centers of our tale. This is not an almanac of 1832, nor is it a biography of Charles Darwin, though both are at the heart of things. It is a look at 1832 with an eye on Darwin and a look at the life and work of Darwin with an eye on 1832 . . . but let's go back to clouds and doors.

It is always possible to adjust history's lens so that any wrinkle becomes a smooth surface; where the more things change, the more they stay the same. The camera shows us what is there, but it has many tricks. There's slow film: transient objects leave no mark. Take a picture of a crowded street and the crowds disappear. No individual is "there" long enough for its image to be recorded. Only the buildings, the street, and sidewalk leave their mark.

There is a photo of a tumbling spring, but the water flows thickly, like treacle. No ripples or eddies, no splashes or sprays of droplets off of rock and boulder. It is what the eye would see if time were slower off the mark.

Adjust our lens again and it is all busy-ness. A mote dances quixotically in a winter's sunbeam, but this only hints at the frenzy that is really there. The molecules of air are rushing at a thousand miles per hour, careening into one another with a violence that makes NASCAR look stately, even sedated. The same incessant motion, the same incessant change is in the crowd that the camera didn't see. Each individual is navigating his or her way through the sidewalk, through the minute, through the day. A young woman, who is deaf, has left the office at the end of the day. She is happy with her work with the Encyclopaedia Britannica Company in downtown Chicago. Her first real job. She is walking down State Street or maybe it's Randolph and a man catches her eye and asks what time it is. She says "4:25" and sees his face change; a look of disgust comes over his face. It was her voice, the strangeness of a deaf person who has spent a lifetime not hearing her own voice. She works hard not to catch anyone's eye after that.

The street is full of such scars. Or appointments filled with possibility: lunch with a new friend, a host of decisions made or postponed. It is Carl Sandburg's city, as in this snippet where the city says to the people:

> I am the woman, the home, the family,
> I get breakfast and pay the rent;
> I telephone the doctor, the milkman, the undertaker;
> I fix the streets
> For your first and your last ride—[1]

It's a matter of your lens, the comings and goings, paying the rent, fixing the street. It is the incessant busy-ness of the city "whistled in a ragtime jig down the sunset." It is the incessant busy-ness of history, that busy-ness which carries yesterday forward to today.

In his rambling commentary *England and the English* (1833), politician and man of letters Lord Bulwer-Lytton observed: "Every age may be called an age of transition . . . but in our age, the transition is visible."[2] It's a matter of one's lens.

The sun rises and the sun sets. Seasons give way to one another. However we might adjust our lens, a unit like the year 1832 must be arbitrary. For sure; yet that does not mean it cannot have been a year apart from other years. At the start of a lovely essay, Stephen Jay Gould tells a story about "Skyline Arch" of Arches National Park and about a pamphlet published by the Park Service. The pamphlet tells us the world changes in imperceptible ways. Rub a canyon wall and hundreds of grains of sand are dislodged. This might seem insignificant, but that is how canyons are formed. Indeed most of the time the pace is slower, the pamphlet goes on to say, but in time you can tear down a mountain or create a canyon, just a few grains at a time. Yet a few pages later in the same pamphlet we learn that in 1940 a large block of stone had fallen from Skyline Arch and it was suddenly twice as large, opening up now to greater vistas.[3] A few grains of sand . . . a massive collapse of rock. Every age may be called an age of transition, but sometimes the transition is visible.

A large block of stone fell in England in 1832—and the young Darwin was struck by his own metaphorical block while off in the unhappy wilds of Tierra del Fuego, a barren cluster of islands off the southern tip of South America. For England as a whole and for Darwin on his own there was a doorway in 1832 and new horizons posing both threat and promise.

THE PROTAGONIST

I have a story to tell about a young man, but when I mention his name, Charles Darwin, he ceases to be a young man and becomes the young Charles Darwin—a very different thing. That difference is really what the story is about. There was a time when Darwin was a lad: I am tempted to say "just a lad," with the potential to become a great scientist. What happened so that a young man, perhaps not so different from many another, became a young Charles Darwin? As it happened, it had a lot to do with events in 1832.

Something happened in 1832, something that constituted a problem Darwin could not let go of. It would lead him across decades to his theory of evolution and on to a theory about the descent of humankind . . . a long lifetime's work.

For students of the history of ideas, Darwin is a "problem shift." Before his work we thought about things one way, afterward we thought about them differently. The gaze of scientist and layperson took in new vistas and new issues appeared in the landscape. But we should not forget that Darwin was also a person, and in 1832 he was young, a recent graduate of Cambridge, and out in the world far from family and friends. It is hard for us to appreciate just how far away he would have felt himself. Unlike today, there were no cell phones, no Internet, and no familiar businesses in universal malls; only an occasional ship carrying news of England and perhaps some mail. The world he met with in his travels would have carried everywhere a very different stamp from life back home, especially in the bitterly cold and relentlessly wet archipelago of Tierra del Fuego.

There is an extensive literature surrounding Darwin, most of it given to Darwin as a problem shift—to the arguments he offered, their various elements and the evidence he pointed to. But what led him to this set of issues in the first place? The pace of evolution is so slow, the changes from one generation to the next so meager that he could not have witnessed evolution and then wondered what had caused it. Evolution was, and is indiscernible, and like the atom it was invented long before it could be discovered.

Consider: atoms are small, so small that you cannot hold one up to the light to see it. You cannot put one on a scale to weigh it, nor point to one in your hand or on a table. It cannot be experienced, and so exactly unlike the New World, it could not have been bumped into on the way to somewhere else. It could not have been discovered; it was invented.

Evolution is not an object. It's an explanatory framework, a way of understanding the history of life. So the analogy with the atom is not with the theory Darwin offered, but with the notion of evolutionary change—that bit of difference from one generation to the next that Darwin posited must be there and would accumulate over time to carry you from pond scum to fish and on to frogs and dogs and people. But what led Darwin to posit these evolutionary bits of change? He could not have seen them, could not have discovered them; why invent them? And still more compellingly, why pursue them for years? Darwin begins his first notebook on evolution in 1837, shortly after he settles in London on his return from the voyage of the *Beagle*.[4] *On the Origin of Species* is not published until late in the 1850s. That's twenty years, a long time to look for the vanishingly small ghosts of departed entities. We will find that the answer lies in Tiera del Fuego, and it requires that we consider Darwin as the person he was. Looking at Darwin with an eye on 1832 is crucial. It gives us the press of a problem, the push behind his work.

The year 1832 also gives us the engine, the analytical machinery that carried his work. This is the other leading aspect of our study: 1832 with an eye on Darwin. As it happened, Darwin comes to the analytical framework that shapes his work in 1832. And there you have it, both motive and means; hence the "making" of Charles Darwin. "Making," by the way, instead of "formation," for example, because "formation" is more definite. It's closed. The deed has been accomplished. But "making" takes you more directly to the moment. It's the act of becoming. It's the way things were at a moment, if we were slower off the mark.

REMEMBRANCE OF THINGS UNKNOWN

We have already borrowed from Carl Sandburg once, let's do so again. In "Good morning, America" he wrote: "History is a living horse laughing at a wooden horse."[5] What a delightful image. A wooden horse is such a feeble imitation. How natural to laugh at it. Just so, the actual past laughs at the feeble imitation of the past fashioned in our efforts to understand. It will laugh at our tale, too.

"History" is a distinctly ambiguous term. Usually, we distinguish what we think about something and the thing itself. Biology is about living things, geology about the earth. But "history" refers to both the past and how we understand it. History is about history. Sandburg pits the quixotic vitality of the real thing, the relentless unfolding of yesterday into today across a myriad of days, against the shadow of history . . . our wooden understanding.

In England in 1832 nothing seemed to be what it had always been. Traditional society had given way to transformation. Successive generations had worked on the lord of the manor's estate, living in the same cottages, working the same soil, toiling the same toil. But no longer. Now vast numbers were "operatives" tending the new machines of industry in Southwark and other parts of London, in Birmingham, Manchester, and Leeds—without tether or community.

Where are we headed? Am I safe? How can I make my way? Such were the pressing questions history addressed, even though it was a mere wooden horse.

At the close of one of his chapters in the *Principles of Geology*, Charles Lyell quotes Niebuhr's *History of Rome*: "He who calls what has vanished back again into being enjoys a bliss like that of creating."[6] But Sandburg trumps Niebuhr and Lyell here, for what they bring into being is but a wooden replica, a Pinocchio who never becomes a child.

Nevertheless, we are drawn to the making of wooden horses, and we respond to the feel of the past with a certain reverence. In the seventeenth and eighteenth centuries gentlemen would have cabinets displaying valued objects, oftentimes objects of history—a Roman coin, a fossil, perhaps a Saxon shield—objects that inspired awe and prompted a story. We stand in an ancient square and in the worn contours of its cobblestones we feel the ages of its busy-ness. We gaze at statuary in a fountain and feel the thirst for both water and art in the countless thousands of souls who would have stood gazing just as we are over its hundreds of years. We pick up a well-worn, dusty text of commentary on a wisdom of a distant past we only know as a footnote—what was once vital and mainstream is now only marginalia—and we feel the weighing of words, the careful appreciation of nuance without knowing its real meaning. This is what Bernard Malamud is pointing to in a story he tells about a man who has worked hard all his life and takes himself from the Bronx to Italy for a year of museum-going and study. He wanders the streets of Rome, captivated by the richness of its past, a richness he feels more than knows . . . this feel of history he calls "the remembrance of things unknown."[7] For sure the history we compose is a poor reflection of the actual complexity of things, yet it resonates with a need to connect with our past, even as we know we cannot really know.

BUCKETS FROM AN ENGLISH SEA

Our world was not made in 1832, but history passed through its streets and a deep mark was made, a scar carved into its busy-ness. A rock may remember a scar for centuries, till weathering wears it away. It is the same with ideas and the feelings they arouse. The intensity fades as edges wear away and scars smooth out. There was an intensity to 1832 both for England and for the young Charles Darwin in the wilds of Tierra del Fuego. If we work at it, perhaps we can bring it back, even though time has long since worn away its edges.

Some of what happened in 1832 was the culmination of long-brewing changes. The reform of Parliament finally occurred after centuries of failed efforts: a few grains of sand . . . a massive collapse of rock. England in 1832 was not scarred by war, but there was violence. People took to the streets. The scent of revolution was in the air. For many that was the odor of fear, fear of the mob. But there was no stand-off at Concord Bridge, no storming of the Bastille. Profound political change took place, but it was by the ballot, of the ballot, and for the ballot. This is where our tale will begin, with political change and the Reform Bill of 1832.

Nor was Darwin scarred by physical violence in 1832, but there was violence for him as well, a violence done to understanding, to deeply held sensibilities. The scent of cruelty was in the air, prompting a discomfort so deep that it shaped a lifetime's work. This is where our tale will end, with the descent of man.

But the question of resolution, of the number of pixels per inch on our screen, remains. In the preface to his fine study *Eminent Victorians*, Lytton Strachey suggested no one would ever write a history of the nineteenth century. There was simply too much information. Certainly one could never hope to write anything like Henry Fynes Clinton's *Fasti Hellenici*. Written in 1826, this was a "spreadsheet" on ancient Greece, chronicling political and military events, artistic and literary creations, scholarship, key details on key figures, and the like, year by year, across hundreds of years from before the first Olympiad in 776 B.C. to the death of Rome's first emperor, Augustus, in 14 A.D. To do this for even one year of the nineteenth century in England would be overwhelming. Strachey suggested the best we could do is "row out over that great ocean of material, and lower down into it, here and there, a little bucket, which will bring up to the light of day some characteristic specimen, from those far depths, to be examined with a careful curiosity."[8]

The image seems apt. If we go along by bucketfuls we can maintain a resolution that allows us to capture the tone of the times. A parliamentary speech, snippets from journals and autobiographical sketches, a sermon, some scholarly articles and manuscripts, these are the stuff of our study . . . all "to be examined with a careful curiosity."

There are many stories to tell. Let's sample from our English sea.

THREE BUCKETS FULL

Plague

We learn among a host of items that there had been an outbreak of cholera in Leeds in 1832 that killed over seven hundred people. Dr. Robert Baker's report to the Board of Health on this epidemic read in part:

> We are of the opinion that the streets in which malignant cholera prevailed most severely, were those in which the drainage was most imperfect; and that the state of the general health of the inhabitants would be greatly improved, and the probability of a future visitation from such malignant epidemics diminished, by a general and efficient system of drainage, sewerage and paving, and the enforcement of better regulations as to the cleanliness of the streets.[9]

The prevailing theory at the time was that cholera was caused by bad air, a miasma—perhaps as we might think of the flu. We can see this very clearly in a book that was also published in 1832 and carried the striking title *The Working-Man's Companion: The Physician.1.Cholera*. This self-help book was written by John Conolly, who had been appointed Professor of the Practice of Medicine at the newly founded University College of London in 1828, where he began to revolutionize the care of the mentally ill. *The Working-Man's Companion* was one of the first titles published by the also recently formed Society for the Diffusion of Useful Knowledge. As the preface explains, the Society was seeking to provide "in a cheap form . . . such plain and useful information relating to Medicine as may be serviceable to the working-class of readers."[10] There had been several recent outbreaks of cholera on the continent and now in Britain. People needed to be alerted to what they could and should do. The *Companion* begins with a primer on the human body, skeleton, musculature, internal organs, and so on, and then proceeds to basic notions about good and ill health, focusing on cholera.

For cholera, as for yellow fever, the plague, and several other diseases, the author explains: "Something hurtful is supposed to be added to the usual air,—something hurtful, but which science has not yet succeeded in detecting. It cannot be seen, or tasted, or touched, or smelt; it has neither palpable substance or colour; but we believe that it exists, because of certain effects which we know not to arise from those parts of the air which we can see and examine."[11] Conolly's study offers the prevailing understanding of cholera before the technology and techniques that established the existence of bacteria and viruses.

Baker's study of Leeds was a crucial shift. Not because of a new theory about the causes of cholera, but because of his focus on the community rather than the individual. The report contains a schedule of the streets, lanes, and alleys where cases of cholera occurred. Rather than any link between the disease and the victims' age, sex, occupation, diet, or even moral character, a common notion at the time, he linked the spread of the disease to sanitation. Baker divided the population of Leeds into two roughly equal portions. In that half where there were sewers, drainage, and pavement, there were 245 cases of cholera. But in the other part of town, lacking such basic amenities, there were 1,203 cases.[12]

It is hard to appreciate how unsanitary were the prevailing conditions. Half of Leeds lived along unpaved streets with no sewers. Commonly dwellings clustered in "yards." The Boot and Shoe Yard, for example, was home to 340 people in 34 units. As many of these houses took in itinerant workers, numbers could swell to over 600. All of these homes were without any plumbing. The "necessary" or toilet was often a wooden screen around a hole in the ground, and sometimes no more than a bucket that could be emptied into a common midden (see figure 1.1).

Later, Baker's views would be incorporated by Edwin Chadwick in his *Report on the Sanitary Conditions of the Labouring Population of Great Britain* of 1842, and they were part of the background for Dr. John Snow's work with a cholera epidemic in London in 1854. Snow was able to link the outbreak to a particular well, thereby demonstrating that cholera was carried in the water. Given the importance of these studies, Baker's work on the outbreak in Leeds in 1832 played a key role in the rise of the discipline of public health.

Murder

Drawing our bucket for a second time, we discover the Anatomy Act of 1832. As Baker's report effectively marks a new and promising chapter, the Anatomy Act brought to a close an old and unhappy chapter. There had long been a tension between the pursuit of medical knowledge and taboos about dissection. Over time a system had evolved whereby the bodies of executed criminals were given to hospitals. The Royal College of Physicians had been allotted up to six corpses by Queen Elizabeth, and a royal grant further allocated up to four to the Company of Barber-Surgeons.[13]

Public executions were relevant here, as the hanged felon forfeited his claim on his own body. As a consequence, executions, already quite the spectacle, often became a battle scene. Francis Place, whom we will come to know shortly, commented: "The whole vagabond population of London, all the thieves, and all the prostitutes, all those who were evil-minded, and some, a comparatively few curious people made up the mob on those brutalizing occasions."[14] The families and friends of the deceased were offended by the ensuing violation of the sanctity of the body. It was enough that life should be taken away. They felt this so strongly that they would often rush the gallows to steal away the body from the authorities. Here is one description from 1740:

> As soon as the poor creatures were half-dead, I was much surprised before such a number of peace officers, to see the populace fall to hauling and pulling the carcasses with so much earnestness, as to occasion several warm reencounters, and broken heads. These were the friends of the person executed . . . and some persons sent by private surgeons to obtain bodies for dissection. The contests between these were fierce and bloody, and frightful to look at.[15]

To add fuel to the proverbial fire, sensibilities changed within the medical community across the eighteenth century. Early in the 1700s the format for dissection had been

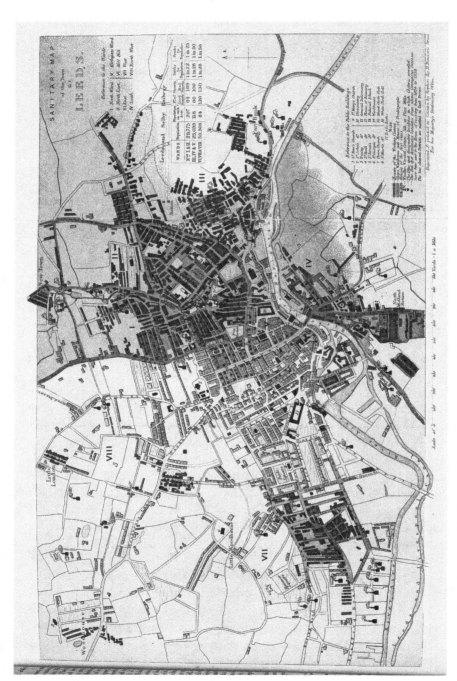

FIGURE 1.1 Robert Baker's map of the cholera epidemic in Leeds, 1833, from E. Chadwick's *Report on the Sanitary Conditions*, 1842. Courtesy of British Library.

a demonstration. A barber-surgeon would perform the incisions, while a professor would address the attending students. To witness was training enough.

The print in figure 1.2 was accompanied by a poem, composed by the Rev. James Townley, a friend of Hogarth's. The first stanza reads:

Behold the Villain's dire disgrace!
Not Death itself can end.
He finds no peaceful Burial-Place,
His breathless Corse, no friend.

Here was the issue in a nutshell—he finds no peace, his corpse, no friend.

FIGURE 1.2   William Hogarth's engraving *Reward of Cruelty*, plate iv in a series of four, 1751. Courtesy of Tate Britain.

Across the century, the direct experience of the dissection came to be valued over a mere witnessing. This put a far greater premium on cadavers. The shortage of corpses in turn meant medical schools did not scrutinize their suppliers too closely. A lucrative trade emerged in body snatching and grave robbing.

In response, night watchmen were hired. Their stalls or stations can still be found in older cemeteries.[16] Grave robbing was bad enough, but the lure of money in hard times led to an even more horrific practice. Likely candidates, such as the down and out, were *murdered* and their bodies sold for cash!

This was the case in the famous Burke–Hare murders in Edinburgh. Burke and Hare murdered sixteen victims in 1828. They were caught and their testimony revealed that they had been supplying a Dr. Robert Knox, a surgeon who regularly lectured on anatomy for a fee. They were paid 10 pounds sterling on average for these corpses.

The scale of the problem can be seen from the 1831 confessions of one London group who admitted to stealing five hundred to a thousand bodies for anatomists, over a twelve-year career. They received 8 to 10 pounds for each cadaver accepted. To try to gauge the value of this amount, a tradesman might make 17 shillings a week. A single cadaver would thus correspond to more than three months of a tradesman's pay.[17]

These practices caused widespread fear and revulsion, and led to a parliamentary investigation. In February of 1832 a bill was brought to the floor. Among those commenting on the bill was Thomas Macaulay. He begins by asking: "What are the evils against which we are attempting to make provision?" "Two especially," he continues, "the practice of Burking, and bad surgery." Macaulay then offers that the poor alone are exposed to these two horrors:

- "What man, in our rank of life, runs the smallest risk of being Burked? That a man has property, that he has connections, that he is likely to be missed and sought for, are circumstances which secure him against the Burker. . . . The more wretched, the more lonely, any human being may be, the more desirable prey is he to these wretches. It is the man, the mere naked man, that they pursue."
- "Again, as to bad surgery; this is, of all evils, the evil by which the rich suffer least, and the poor most. . . . The higher orders in England will always be able to procure the best medical assistance. Who suffers by the bad state of the Russian school of surgery? The Emperor Nicholas? By no means. The whole evil falls on the peasantry. If the education of a surgeon should become very extensive, if the fees of surgeons should consequently rise, if the supply of regular surgeons should diminish, the sufferers would be, not the rich, but the poor in our country villages, who would again be left to mountebanks, and barbers, and old women, and charms and quack medicines."

And Macaulay concludes: "I think this a bill which tends to the good of the people, and which tends especially to the good of the poor. Therefore I support it. If it is unpopular,

I am sorry for it. But I shall cheerfully take my share of its unpopularity. For such, I am convinced, ought to be the conduct of one whose object it is, not to flatter the people, but to serve them."[18]

The bill passed, providing medical schools with an adequate supply of cadavers of those who were already dead.

Art

Our third bucket comes from the world of art. Constable and Turner were leading painters of the day, and both were exhibiting paintings at the gallery of the Royal Academy of Art in 1832. An intriguing practice at the time was that members of the Academy were invited to set their paintings up and continue to work on them in the company of fellow members and the public prior to the formal opening of the exhibition. During this period Turner, it seems, added a splash of red paint to one of his seascapes, and then left without saying a word. Evidently, he felt the painting needed more focus.

Constable, struck by Turner's deft touch, remarked: "He has been here and fired a gun."[19]

Turner is a most intriguing character. Across a long life (1775–1851), he painted and sketched a host of works, perhaps as many as fifty thousand. He began as a lad in his father's barber shop, selling portraits of the customers. He trained hard, working as a topographical draughtsman for an engraver and later as a copyist of other painters' works. This early training may have been the key, for he was able to immerse himself in both the careful delineation of beautiful and picturesque views and the more imaginative use of landscapes.[20]

Turner is one of those figures of whom it is often said he was ahead of his time. And certainly his mature work evokes atmosphere and uses light in ways that speak to the work of the Impressionists decades later at the end of the century and beyond, qualities we can see in figure 1.3. Yet we should not see him as out of sync with his day. His art resonated with the times. He was certainly commercially successful and was a leading member of the Royal Academy. But he also pushed his art. How should we understand this push?

We may start with the notion of authenticity of *place*, acknowledging his training as a young man in preparing detailed views of particular places at the heart of topographical works.[21] On top of this, we can see Turner working to evoke the authenticity of the *moment* as well, with all the vagaries of light on a given day. The topographical painter brought the clarity of studio light to the landscape. Turner, and later the Impressionists, went further to render authentically the light of nature. His images became shrouded in mists and fogs, set at dawns and dusks, even in the middle of the night, midst storms and squalls. There is detail and structure, but it is now more suggested than delineated, and so it required different techniques to attract the eye and hold the imagination. When Turner added that splash of red, we can see him satisfying the demands of an emerging aesthetic that revolved around authenticity.[22]

FIGURE 1.3 J. M. W. Turner's *Helveotsluys*, 1832. Courtesy of Tokyo Fuji Art Museum, Tokyo, Japan/ Bridgeman Images.

Turner, we may add, returned to the painting a second time and reworked the patch of red into a buoy. He had first to meet the demands of the aesthetic. Only afterward did he fashion it into something that might plausibly be a part of the scene. Here was transition made most vibrantly visible.

MORE THAN BUCKETS

Here, then, are three buckets from the English sea in 1832. Such a random sampling of items may include many a good story, but it will not take us where we want to go. We have our eyes on Darwin and on the influences that pushed and pulled him about. Casting our bucket widely across this English sea, we will sample the waters of politics, classical history, and geology, looking for how people sought solutions and what strategies they leaned upon. We will find deep commonalities, commonalities that suggest two distinct and far-reaching structures of analysis that resonated with 1832 as individuals sought to meet the needs of a society confronting profound changes.

As a unit of time, 1832 is crisp and well-defined: 12 months, 52 weeks, 366 days (it was a leap year). Yet as we tease out the analytical influences at play in this moment, we will find that the geographical reach is extraordinary. The context for the young Darwin, fresh out of university and ready to make his mark, to assert his place in the scheme of things, takes in the books he was reading while on board the *Beagle*, such as Alexander von Humboldt's *Personal Narrative*, the account of his travels into the tropics of South America. As he waited for the *Beagle* to set sail, Darwin wrote his mentor, John Henslow, Professor of Botany at Cambridge, describing his excitement at the prospect of going to Tenerife in the Canary Islands and visiting the sites and vistas that Humboldt had so vividly portrayed. As we sample buckets, follow leads, and trace lines of reasoning we will find ourselves visiting sites and conjuring vistas across the globe: a shop for the making of leather britches in London's East End, and a bookbinder's off Piccadilly Circus; the beach at Tyneside; an astronomical observatory in Cape Town, South Africa; the Chapel at Trinity College; the hill country of North Wales; a bishop's palace not far away; and the bleak rocky islands of Tiera del Fuego. All of these are part of the English sea we will sample.

PUSHING TOWARD THE END

This aspect of our study focuses on the resources individuals drew upon as they coped with the challenges before them. "Resources" is too bland a term. It is a lot more active than that—there is a *push* involved.

The word "push" has its own tale, one that takes us all the way back to ancient Greece and the emergence of the sciences. By tradition, the first book on science, *On the Nature of Things*, was composed by Anaximander (about 590 B.C.). Giorgio de Santillana explains that actually it was *On the Physis*, where the Greek "physis" is the root of such English words as "physics" and "physician" and also of the verb "to push." "Physis" was originally an agricultural term, referring to the push of a seedling as it broke the soil and asserted itself in the scheme of things. Though this is a nicely naturalist sort of phenomenon, why was it seen as capturing the qualities Anaximander and his fellow "scientists" were after? What is it about a seedling asserting itself that captured this new arena, "science"? Things begin to fall in place once we realize that there was a corresponding Latin term, "natura," that was also originally about the push of a seedling. Though we tend to think of "nature" as what is out there, in fact it is very much more about what is within.[23] We preserve this internal, assertive quality when we speak of someone's nature, something innate that makes them what they are.

What had interested the ancient Greeks was the push within things to make them behave the way they did: their nature. Our interests lie chiefly in the thinking behind the events and scholarship of 1832 that led to their unfolding the way they did. What was the nature of the push behind things at this moment, a moment when change was all about and you couldn't simply go on about your busy-ness, when you had to stop and figure things out? As we examine our buckets, patterns will emerge and these patterns will enable us to appreciate the making of the young Darwin as he coped with a profound challenge raised during his voyage on the *Beagle*.

We will start with the Reform Bill, the flagship of the changes of this era. Then we will move from Parliament to scholarship, but without leaving worldly affairs. The image of the scholar is often otherworldly, and no doubt there is an escape from the press of everyday life in concerning yourself with the nuance of an interpretation or a variation in a standard lab procedure. But scholarship need not be indifferent to the times.[24] The central cast of figures in our study presents an intriguing array of professions, including a tailor, a professor, a banker, and a bishop. Each was an intellectual, and each confronted the deepest problems England faced in 1832—its institutions and its core cultural values.

The end of our study, then, is to sample the waters of politics, science, and letters in England, focusing on the year 1832, with our eye clearly fixed on the events and scholarship of this year and further, how they led the young Darwin to tackle the task of forging a powerful understanding of the history of life. In 1832 transition was palpable: there was political unrest, with riots in Bristol and the streets of London resounding with marching, charging feet, and there were also powerful new ideas in scholarship—in the lines and lessons that would be drawn from myth and from fossil in the effort to make sense of things anew.

And though I hardly expect to avoid the laughter of the living horse, I do hope to set before you a significant and engaging tale of the effort to cope in an age where the air was charged with the scent of change.

NOTES

1. "The Windy City," in Sandburg, *Complete Poems*, 278.
2. Bulwer-Lytton, *England and the English*, vol. ii, 108.
3. Gould, *The Panda's Thumb*, 194–95.
4. Darwin's first notebook on transmutation is notebook "b." Other early notebooks on evolution: "red," "c," and "m." These jottings related to evolution are from 1837–1838. See Darwin, *Charles Darwin's Notebooks, 1836–1844*.
5. Sandburg, *Complete Poems*, 326.
6. Niebuhr, *History of Rome*, 5. Cited in Lyell, *Principles*, vol. i, 74. There is an intriguing issue here. As it happens, I have studied with three scholars who sought a fictional or hypothetical setting for their critical scholarship. Imre Lakatos wrote a brilliant piece, *Proofs and Refutations* (1976), where he brought mathematicians together in an imaginary classroom for a discussion of a particular hypothesis. In the course of their discussion they each gave voice to what they had written on the topic, but more richly Lakatos was able to use their exchange to tease out a powerful set of notions about the growth of mathematical ideas. Russell MacCormmach wrote a biography of an imaginary German physicist toward the end of the nineteenth century, drawing attention to both the state of the discipline and the many pressures within the German scholarly world: *Night Thoughts of a Classical Physicist* (1982). Lastly, Robert Bakker used his many paleontological studies to pull together a memoir in the voice of a velociraptor. *Raptor Red* (1985) was an extraordinary piece, as were the other two. Each was historical and each author presumably could feel Sandburg's horse laughing in the background. They wanted, somehow, to bring their subject more fully to life.
7. Malamud, *The Magic Barrel*, 162.
8. Strachey, *Eminent Victorians*, vii.
9. Baker, *Report*; Chadwick, *Report on the Sanitary Conditions of the Labouring Population*; and Snow, *On the Mode of Communication of Cholera*.
10. Conolly, *Working-Man's Companion*, 7.
11. *Ibid.*, 43.

12. Baker, *Report*, 14. See also Brooke, "A Tidal Wave of Disease," 41–44; and also Discovering Leeds, section on poverty and riches, at www.leodis.net.

13. Linebaugh, "The Tyburn Riot Against the Surgeons," 70–71. Linebaugh cites both a petition recorded in the *Commons' Journals* and *The Annals of the Barber-Surgeons of London*, as well as R. Shyrock's *The Development of Modern Medicine*.

14. Place, British Museum, Add. Mss 27,826 Place Collection "Grossness," fo. 107; cited in Linebaugh, "The Tyburn Riot," 68.

15. Linebaugh, "The Tyburn Riot," 81.

16. Vale, *Curiosities of Town and Countryside*. Opposite p. 14 is a photograph: "The Regency Watchman's shelter, Old Wanstead Church, Essex. A reminder of body-snatching days."

17. On the Burke–Hare murders see Thomas, *The Doctor and the Devils*; Rankin, *The Falls*. One pound sterling from 1830 is roughly equivalent to $100 in 2010 dollars: see measureworth.com. Finally, there is a curious link between our first two buckets: Dr. Baker of Leeds had been tried for paying for a snatched corpse! He was acquitted because it was to advance knowledge. See Rosenhek, "Invasion of the Body Snatchers."

18. Macaulay, *Miscellaneous Writings and Speeches of Lord Macaulay*, vol. iv, 425–26.

19. Jones, "Turner and Constable Exhibitions," *Guardian*, August 24, 2014. See also *Chambers's Journal of Popular Literature, Science, and Art*, 4th series #234(1868)395.

20. Butlin, "J. M. W. Turner," 9.

21. Ball, *Science of Aspects*; Ladd, *Victorian Morality of Art*; Ruskin, *Modern Painters*, vols. i and v.

22. Consider another painting by Turner in 1832, Venice. It is a view late in the evening and a large building, perhaps a grand home or a warehouse, is really a large block of shadow marked by a narrow slash of light, along with its reflection in the canal. The effect is more commanding and I find this painting more compelling, but it was a less public event.

23. De Santillana, *Origins of Scientific Thought*, 27.

24. See de Santillana, *The Crime of Galileo*. It is not always the case that scholars will so explicitly link their work to affairs of the day; nevertheless, there are many instances where they have clearly responded to the issues of the day, be it philosophy, as in Karl Popper's *The Open Society and its Enemies* (1945), a reading of the history of philosophy written during World War II and directed against fascism; or psychology, as in Robert Coles's *The Moral Intelligence of Children* (1997) about racism; or literary criticism, as in Gilbert Murray's *Aeschylus* (1940), aimed at academic freedom.

2

OUTRAGEOUS TO MORALITY, PERNICIOUS TO GOVERNMENT

Parliament has been around for a long time. Back as early as the seventh century, well before William the Conqueror and the Norman invasion, kings would on occasion seek the advice of the landed elite, meeting with a council of leading barons in a meadow at Runnymede, the same site where the barons would force King John to sign the Magna Carta in 1215: a key moment in the long-standing struggle between the power of the king and that of the aristocracy. In 1295 King Edward the first called leading barons and prelates for advice, but this time he also included elected representatives of smaller rural landowners and of townsfolk. Over the generations this notion of both landed elite and elected representatives would go through various transformations, but after the turmoil of the English Civil War and the return of the monarchy in the seventeenth century things so settled into place that there would be little change through to 1832.

The system of representation in the House of Commons was fairly elaborate. There were 658 seats, 514 of those from England and Wales. These were broadly of two sorts. Most counties had two representatives, the idea being that these members reflected the interest of the landed gentry and elite. In addition, there were members who represented boroughs or towns and their mercantile and trading interests. We should note that the lord of the manor had pervasive influence over adjoining communities, and so MPs from particular towns would be beholden to the local lord. This is further underlined when we note that voting was a public pronouncement and not by secret ballot. It would have taken considerable courage to vote for someone other than the candidate favored by the lord of the manor.

The House of Commons had not been envisioned as representing the population numerically, but rather by leading interests. The bishops of the Church sat in the House of Lords and both Oxford and Cambridge universities had their own MPs, for instance. This was an intriguing approach that saw Parliament as an arena for the core institutions of society to safeguard their interests. Though the House of Commons today is based upon population, we learn from a charming collection of letters to the editor of the *London Times* that of the 615 MPs after the election of 1935, twelve represented universities—a trailing residue of the old ways that persisted until 1950.[1]

Returning to the state of things on the eve of the Reform Bill in 1832, the system had accumulated a number of peculiarities. Dunwich, for example, which had been a leading town in medieval East Anglia, had pretty much fallen into the sea late in the thirteenth century, but still had a seat in parliament. No one lived in Old Sarum any-more either. Indeed, no one had lived there for a long time. Folks moved out from this hill top settlement down to the banks of the Avon River early in the thirteenth century. Over time, New Sarum became the cathedral city of Salisbury, but Old Sarum retained its member of parliament. Such boroughs were called "pocket boroughs" because the representative in Parliament for that seat was in the pocket of some member of the landed elite. Further, since no new borough seats had been added after 1661, the growth of such industrial centers as Manchester, Leeds, Sheffield, and Birmingham was not represented. The system had utterly failed to keep up with the dramatic changes of the Industrial Revolution.

REFORM AND CHARLES JAMES FOX

In the various debates across the end of the eighteenth and the beginning of the nine-teenth centuries, the most formidable voice calling for the reform of parliament was that of the great Whig leader Sir Charles James Fox.

The son of the Duke of Holland was a fine teller of tales, a notorious carouser and gambler, and a gentleman of means. Sam Johnson was a creature of very different hab-its, but he made an exception with Sir Charles. Fox was an early member of Johnson's Literary Club, a small group of ten or so who met frequently for good food and good talk. It says a lot about Sir Charles that he found favor in the eyes of Sam Johnson. In a lovely passage on Fox, the historian G. M. Trevelyan sums up his character with these words: "Oratory at its highest, politics at its keenest, long days of tramping after partridges, village cricket, endless talk as good as ever was talked, and a passion for Greek, Latin, Italian and English poetry and history—all these, and alas also the mad-ness of the gambler, Fox had enjoyed and had shared with innumerable friends who loved him." And, Trevelyan adds: "Nor had he been less happy during the long wet day at Holkham which he spent sitting under a hedge, regardless of the rain, making friends with a ploughman who explained to him the mystery of the culture of turnips."[2] This hints at the fuller measure of the man, but we can round it out still more if we turn to another matter altogether.

It was 1968 or 1969 when the *Washington Post* ran an op-ed piece written by Sir Charles James Fox. It was extracted from a speech Fox had given nearly 170 years ear-lier on the evening of February 3, 1800, but the *Post* piece simply gave him the byline with no further explanation. Here is the gist of it: William Pitt, still Prime Minister, had defended his prolonging of the war against Napoleon as he sought an honorable peace—or was that Nixon's phrase? Hard to keep these things straight sometimes. It was clear to everyone it was a war we wanted no part of any more, and so the gov-ernment was beating a retreat but not a hasty one lest anyone suspect it had been a mistake in the first place. So, said Pitt, the war had been suspended. It was not really a war at all, but a holding action, and with this Sir Charles, the leader of the opposition in the House of Commons, rose to his feet and addressed the divided house, inviting them to put themselves in the field of battle. Imagine what it means to pause: "But if a man were present now at a field of slaughter, and were to inquire for what they were

fighting—'Fighting!' would be the answer; 'they are not fighting, they are *pausing*.'"
And, Sir Charles continued: "'Why is that man expiring? Why is that other writhing
with agony?' . . . The answer must be, 'You are quite wrong, sir; you deceive yourself—
they are not fighting—do not disturb them—they are merely pausing!—this man is not
expiring with agony—that man is not dead—he is only pausing! . . . All that you see, sir,
is nothing like fighting—there is no harm, nor cruelty, nor bloodshed in it whatever—
it is nothing more than *a political pause*.'"[3]

A bullet fired in a holding action would shed blood as readily as any fired in war,
whether it was the continental campaigns of 1798 or Vietnam late in the 1960s.

Charles James Fox was the parliamentary leader of the Whigs in the House of
Commons for a long time, urging not only a true end to the war with France, but a
host of reforms that would transform the character of the government. He sought to
recast English politics, moving its center closer to the people.

There had been a number of efforts to reform parliament over the years, but we
may pick things up on the evening of May 26, 1797, when Fox addressed his fellow
members of the House of Commons in support of Grey's second effort to initiate parliamentary reform. "The whole of this system, as it is now carried on," he urged, "is as
outrageous to morality as it is pernicious to government."[4] To illustrate the corruption
of the system, Fox referred to a member selling a seat in parliament for four or five
thousand pounds, and then on a later occasion moving to send a poor soul from his
district to prison for having sold his vote for less than two pounds. But an even greater
sin was the way the existing system denied the people a voice in government. Fox then
recalled events some twenty years earlier, when the government had lost the confidence of the people who no longer supported the war against the American colonies.
An election was held. Despite the strength of anti-war sentiment, only three or four
seats were gained by the opposition.

Parliament's systemic insensitivity to the will of the people was a serious matter and
Fox's most prescient argument. He described at length more recent events in Ireland,
where the number of radicals had grown dramatically. In 1791 there were, perhaps,
ten thousand men prepared for violence. It was estimated that this number had now
increased tenfold because the government had failed to support moderate elements. As
he closed his speech, Fox asked what should be gained by passing reform. He answered:

> I think we shall gain at least the chance of warding off the evil of confusion, grow-
> ing out of accumulated discontent. I think we shall save ourselves from the evil
> that has fallen upon Ireland. I think we shall satisfy the moderate, and take even
> from the violent if any such there be, the power of increasing their numbers and of
> making converts to their schemes.[5]

For sure, the threat of violence would enshroud the events of 1832, like gray skies
before a storm, but as we shall see, it was more complicated than one might have
thought. A key party of reformers had inherited Fox's view that reform was the best
way to enhance political order and replace the anger and confusion which grows out of
discontent with a more positive civic spirit.

The division that evening was ninety-three votes in support of Grey's motion and 258 opposed. That parliament would reform itself, that power would graciously acknowledge that it did not comply with principle, certainly seemed unlikely. And so things stood as they had for centuries. Fox died long before the Reform Bill passed, but he had by then passed his discontent and his clarity on to a new generation.

"WHERE THE EVIL IS THUS IMPERFECTLY CONCEIVED"

England never had a French Revolution. The French ruling elite was either executed or exiled in the wake of events at the close of the eighteenth century. And though there were oscillations politically in the succeeding decades, the grip of aristocratic families had been broken.

England stands in contrast. It may be that England's wealth is no longer as concentrated in the holdings of her most ancient aristocratic families, but whatever changes there have been to their fortunes have been the result of the ebb and flow of "ordinary" history, as opposed to the extraordinary proportions of revolution.

Why then is the transition in the England of his day so visible for Bulwer-Lytton? If England did not have a French revolution, it nevertheless had a revolution of its own—the Industrial Revolution. The changes wrought by the development of the factory system and the mass production of cloth and what-have-you transformed virtually every aspect of life. Take the making of straight pins.

One of the first stories Adam Smith tells in the *Wealth of Nations* (1776) is about straight pins. Back before the factory system, the making of straight pins had been a cottage industry. A "jobber" had a circuit. He would deliver a roll of steel wire to workers in cottages across the county.

These workers would then cold-draw the wire. That is, in order to reduce the diameter of the wire, they would draw it through successively smaller holes in a frame. (The same thing was done for piano and harpsichord "strings.") After the wire had been thinned, you would snip it, and lay it over a notch in an anvil and hammer it to produce a point. Several such operations were necessary to make the head and then to whiten the pin. Finally, you would pack it in a roll of paper. All in all some eighteen distinct operations were necessary. The typical worker could make a handful of pins in a day, fewer than twenty; bringing together a small group of ten workers, however, so that each could specialize in a few tasks, increased production to over 48,000 pins a day.[6] Even without the steam engine to drive conveyor belts, the division of labor was an engine of extraordinary productivity.

Industrialization transformed far more than the baubles, bangles, and bright shiny beads you could buy at the store. It changed how people worked and where. The latter half of the eighteenth century saw the growth of several major industrial centers. London reached a population of a million; in addition Birmingham, Manchester, Leeds, Sheffield, and other towns grew markedly. Manchester, for example, had a population of approximately 15,000 in 1760 and more than a quarter of a million in 1831, and in the decade from 1821 to 1831 the population of both Liverpool and Manchester would grow by 40 percent.[7]

Just as factories produced steel rails and cloth, so too did they manufacture slums with cholera and other maladies. Much of England's population shifted from the estates of the landed elite to factories and providing housing was a real test for the economy. Homes needed to be built on a huge scale for a population that would not be able to purchase them. The result was a system of company towns or developments, most often featuring "back-to-backs." Imagine a block of row or terrace houses with two long rows of homes separated by small gardens in the back and an alley between them. Now get rid of the gardens and alley. The homes were back-to-back. Two rooms down, two rooms up, and your back wall was their back wall. The same efficiency that marked the new factory workplace had been brought to workers' homes. And it worked, except for the fact that such density allowed for the rampant spread of disease, as witnessed in the deadly outbreak of cholera in Leeds in 1832 that was our first bucket.

James Kay makes this point in his pamphlet *The Moral and Physical Conditions of the Working Classes*, a study of Manchester in 1832 which echoes Baker's study of Leeds in many ways.[8] Manchester set up fourteen boards of health across the city with inspectors who examined the city, street by street, taking inventory of conditions and surveying the inhabitants. The poor were generally housed in back-to-backs, which meant they had no yards, no privies, and no receptacles for trash: "Consequently the narrow, unpaved streets, in which mud and water stagnate, become the common receptacles of offal and ordure." Such are the conditions that coupled with the exhausting work of the operative in a factory constituted a "predisposition to contagious disease."[9]

In 1831, John Stuart Mill began a series of articles titled "The Spirit of the Age." Here are a few snippets from his effort to capture the tenor of the times:

> The conviction is already not far from being universal, that the times are pregnant with change; and that the nineteenth century will be known to posterity as the era of one of the greatest revolutions of which history has preserved the remembrance, in the human mind, and in the whole constitution of human society. . . .
>
> It is felt that men are henceforth to be held together by new ties, and separated by new barriers; for the ancient bonds will now no longer unite, nor the ancient boundaries confine. . . .
>
> The first of the leading peculiarities of the present age is, that it is an age of transition. Mankind have outgrown old institutions and old doctrines, and have not yet acquired new ones. When we say outgrown, we intend to prejudge nothing. A man may not be either better or happier at six-and-twenty, than he was at six years of age: but the same jacket which fitted him then, will not fit him now. . . .
>
> Worldly power must pass from the hands of the stationary part of mankind into those of the progressive part. There must be a moral and social revolution, which shall, indeed, take away no men's lives or property, but which shall leave to no man one fraction of unearned distinction or unearned importance.[10]

Mill's broad observations were complemented in 1831 by an influential pamphlet on parliamentary reform by his friend George Grote. Grote was a banker and had already earned his reformer's credentials by this time. An active member of the circle variously called Benthamians, Utilitarians, or Philosophic Radicals, Grote had among other projects taken a leading role in the formation of the University of London. We will come to know him well, as we take up his *History of Greece* later in our study.

Grote's pamphlet does not dwell on the outrageous inequalities of pocket boroughs like Old Sarum in contrast to large industrial cities like Manchester and Birmingham with only token representation, though he does note a telling observation Lord Grey had made back in 1793. The young Charles Grey, who would become the second Earl Grey, had been only twenty-two when he entered Parliament in 1786, having only recently graduated from Cambridge. He quickly became one of Charles James Fox's most trusted lieutenants and in 1793 he introduced a bill for the reform of Parliament. Grote goes back to Grey's address where Grey had noted that fewer than two hundred families, partly Peer and partly Commoner, return a majority of Parliament.[11] Grote then added regarding pocket boroughs: "Such abuses are indeed indefensible; but they ought to be attacked, not as vicious excrescences on a system sound in the main, but as symptoms, rather gross and magnified, of widespread internal corruption."[12] This is a critical shift. It is worth following Grote to see where he takes it.

Grote is confident that reform will be secured, but he is concerned that it achieve the desired end. In the measured cadence of his Victorian prose, he wrote: "Where the evil is thus imperfectly conceived, the remedies demanded are likely to be equally incomplete and superficial."[13] And so he takes the conversation beneath the surface, in order to more perfectly understand the evils of the system. Grote takes his analysis back to the Glorious Revolution of the late seventeenth century where Parliament united with popular sentiment against their common foe, the Stuart monarchy. Since that time Parliament had traded on its supposed common interest with the people, but this had forever been false. Those who see Parliament as the guarantors of the security of the people have "overlooked the fact that elections by the people were a pure fiction: that the persons who elected formed only a fraction of the people; and that to this electoral fraction, in the last resort, all the security was to be traced."[14]

We may add this note underlying Grote's point from Antonia Fraser's recent study of the Reform Bill: "Where elections were concerned, just over 3 per cent of the population voted in 1830, some 400,000-odd people out of a population of approximately sixteen million: all were, of course, male."[15]

Grote sees through the gloss that presents MPs as abstractions in touch with the will of the people that mysteriously arises from the practice of England's government. They were, in fact, either members of the ruling aristocracy themselves or were deeply beholden to them. Consequently, Grote links parliamentary reform to the expansion of the electorate, for that is the only reform that promises a harmony of the actions of Parliament and the needs of the people.

Subsequent events show that Grote's shift from rectifying the glaring abuses of parliamentary districting to basic matters of representation was right on target. Later reform bills, after that of 1832, addressed precisely this matter of the expansion of the electorate.

In order for it to be a government for the people, it would have to be a government elected by the people.

FRANCIS PLACE AND THE MEN IN RAGGED COATS

Though England never had a French Revolution, it came close in 1832, and at the center of things was agitation for the Reform Bill. Let's turn from leaders in parliament like Grey and Fox to take up the story of a working man, Francis Place. Place was born in London in 1771. He trained in the craft of making leather breeches and in his early twenties, newly married with an infant son, he began to work in earnest, only to find himself in the middle of labor strife. And though he had not sought a strike, he decided to make the best of things, helping, for example, to write out a statement of grievances. The strike failed. Place's story, already a story of hardship, now took a horrific turn. He was blackballed because he was seen as one of the leaders of the strike. Unable to work, he and his wife faced starvation and lost their son to smallpox.

As hard as it is to fathom the despair that comes from such events, their plight was not far from common among the working classes. Thirty years later, when he had become a wealthy man, Place wrote movingly of this period in his life: how "the disappointments are more than can be steadily met; and men give up in despair, become reckless, and after a life of poverty, end their days prematurely in misery." He went on to talk about the distinctive way such misfortune hits "the better sort of persons . . . who have set their hearts on bettering their condition," reflecting on how "disappointment preys on them . . . hope leaves them . . . their hearts sink as toil becomes useless." Place then noted the overlay of class on all this, writing "it is not the habit of men to care for others beneath them in rank."[16]

Place turned his life around by virtue of extraordinary discipline. The failure of the strike and the suffering it had caused underlined the value of organizing. He worked with carpenters, plumbers, and other tradesmen, helping them to come together effectively. He also did whatever piecework in tailoring he could secure. And, strikingly, he read—at least two or three hours an evening. He borrowed books wherever he could: books by Hume, Locke, Adam Smith, Blackstone. In time, he ventured to set up a shop for himself and was remarkably successful. As a breeches-maker in 1793 he had earned 17 shillings a week, approximately 45 pounds a year. In 1817 when he turned the shop over to his eldest son, he was earning some three thousand pounds per year.

Where does such strength come from? Place has told us it is not from hardship—for hardship preys on people and drives hope away. It was the injustice of it all and the extraordinary disproportion of the consequences. After he had earned a more comfortable standing he did not boast of being a self-made man. He did not rush about with Falstaffian exuberance, or carry himself with the bombast of a man bigger than life, but quietly and steadily he determined to make life less harsh for the working poor.

He became interested in politics, but not as a candidate. He extended his organizing from trades groups to politics. In 1807, while Place was still a tailor, there were parliamentary elections. Francis Burdett, a popular figure in London, was willing to stand for election, but not to spend any more of his substantial funds in the effort. Place and a few others proceeded to organize for him, canvassing in earnest. Burdett was elected and the small group that had worked for him, the Westminster Committee, became the district's central political authority. And so began Place's second career.

Throwing himself into public affairs, Place quickly expanded his work and his contacts. He came to know James Mill well, and through Mill Jeremy Bentham. Place visited Bentham regularly, becoming one of the central figures within the Benthamian circle. Though a practical man of affairs, Place was keen on the exchange of ideas within the circle. One reflection of this is the role he played in the preparation of several of Bentham's notes and manuscripts for publication. This was far more than a matter of sorting out details. Bentham wrote virtually every day of his long life on a range of topics. By 1820 he had accumulated a host of manuscripts, often including several different approaches to the same topic. John Stuart Mill reduced a large number of these to a five-volume work on legal practices, *Rationale of Judicial Evidence*; George Grote wrote a work on natural religion from some 1,500 pages of manuscript; Place worked on several of these manuscripts, including one on schooling, *Chrestomathia*, and another titled *Not Paul, but Jesus*, and he assisted with a third, *A Handbook of Political Fallacies*.

The *Handbook* was quickly and favorably reviewed in the *Edinburgh Review* by the editor, Sydney Smith, and the review clues us in on the matter at heart here in a delightful way: "Whether it is necessary that there should be a middleman between the cultivator and the possessor, learned economists have doubted. But neither gods, nor men, nor booksellers can doubt the necessity of a middleman between Mr. Bentham and the public." The review goes on: "Mr. Bentham is long; Mr. Bentham is occasionally involved and obscure; Mr. Bentham invents new and alarming expressions; Mr. Bentham loves division and subdivision—and he loves method itself, more than its consequences." And the passage concludes: "Those only therefore who know his originality, his knowledge, his vigor, and his boldness, will recur to the works themselves. The great mass of readers will not purchase improvement at so dear a rate, but will choose rather to become acquainted with Mr. Bentham through the medium of Reviews—after that eminent philosopher has been washed, trimmed, shaved, and forced into clean linen."[17]

So one of Place's activities was as "butler" to Bentham's writings, along with other Benthamian projects and everything else he was doing. Place would come to be relied upon as ex officio staff for a number of Members of Parliament, often with regard to labor and tradesmen matters. He was a unique figure in this class society. He had solid footing within the working classes. He had been there, and there is no doubt but that he always remembered being a breeches maker and how powerless he had been to secure work. Since those days, he had also come to have a rich array of political connections with the ruling classes through his electoral organizing, and he was sensitive to the deeper character of political and economic matters. For example, in a two year stretch in the 1820s he worked toward the repeal of various laws regulating such trades as curriers, tanners, and hackney coachmen. He devised schemes to improve the administration of finances, of laws pertaining to debtors and creditors, and crimes at sea. Meanwhile he also supported the formation of mechanics' institutes and helped to negotiate various labor disputes. A charming article in *European Magazine* in 1826 captures Place's reach and his influence with a line borrowed from Archimedes: "Give me *place* for my fulcrum and I'll move the world."[18]

Perhaps Place's single most significant accomplishment across this period was the repeal of the Combination Acts, laws prohibiting the formation of unions or collectives for various trades. He wrote letters to trade societies and newspapers, became involved in disputes, and worked steadily to educate the public. In a letter to Sir John Cam Hobhouse shortly before he, Hobhouse, became MP for Westminster, Place looked forward to a parliamentary committee on labor and wages convinced the repeal of the Combination Laws was at hand, and further that it "would make thousands of reformers among the master tradesmen and manufacturers."[19]

Place's regard for the making of moderates is a recurrent theme, echoing the observations of Fox in his speech for reform back in 1797. Again and again we find Place working to steer public opinion toward reform without revolution. A notable example can be seen in the events of 1819, following the Peterloo Massacre. In July of that year there had been a large gathering in St. Peter's Fields, Manchester in support of parliamentary reform. Manchester, you will recall, had no representation in parliament, though it was already a city of two hundred thousand. Though a peaceful gathering, the authorities had issued a proclamation against seditious meetings. The authorities moved to arrest the speakers, and within moments the military charged to disperse the crowd. Eleven spectators were killed and hundreds wounded.[20]

Reformers across the country were alarmed, fearing that England was losing its cherished right to hold public meetings. There was a desire right away to rally in London, but Place urged caution and the march was not held until September 2—a gathering of one hundred thousand. All "at the risk of military execution and under the musquetry of the household army," wrote Hobhouse.[21] What an extraordinary event this march must have been.

Afterward Place picked up on a phrase in a letter to a friend: "You say, 'the men in ragged coats have proved by their conduct and their resolutions that they understand the business they are about.' This is very true." Until very lately one might have feared a gathering of five hundred, he continued: "Now 100,000 people may be collected together and no riot ensue, and why? . . . the people have an object, the pursuit of which gives them importance in their own eyes, elevates them in their own opinion, and thus it is that the very individuals that would have been the leaders of the riot are the keepers of the peace."[22]

MAKING REFORM HAPPEN

1830

We come now to the events surrounding the passage of the Reform Bill. We may take our beginning with Lord John Russell's proposal in February of 1830 to enfranchise three industrial cities: Leeds, Birmingham, and Manchester. Yet even this modest extension of parliamentary seats was defeated (180 to 140). With the death of King George IV in June of that year new parliamentary elections were called for. At this time, events in Paris had captured the imagination. There had been a coup and a constitutional monarchy under Louis Philippe had replaced the despotism of Charles X. "This new Revolution produced a very extraordinary effect on the middle classes," Place wrote. "Every one was glorified with the courage, the humanity, and the honesty of the Parisians, and the common people became eagerly desirous to prove they too were brave, humane, and honest."[23] The Tory majority in the House of Commons was sharply reduced in this election, but nonetheless they held a majority and the Duke of Wellington formed a government.

Even so, there was a difference. The system had lost its credibility. Place wrote: "There never can be a *strong* government again in England until there has been a change in its very form."[24]

The Duke's government fell to divisions within the Tory party, and in November of 1830 Earl Grey formed a Whig government with Lord John Russell as the leader in the House of Commons. The press for reform grew steadily with political unions in the towns and fires in the countryside. The faculty at Cambridge stopped speaking to those on the other side of the issue, and Place observed that "their steady perseverance, their activity and skill, astounded the enemies of reform."[25]

1831

On March 1, 1831 a reform bill was introduced, generating widespread excitement. It was expected the bill might impose new penalties for bribery, and replace such pocket boroughs as Old Sarum with seats for the likes of Leeds. The Tories were stunned by the scale of the new bill. It dropped sixty of the smallest boroughs, extended the representation of the larger cities, and broadened the franchise to a significantly larger proportion of the population. The Victorian historian Thomas Escott tells us in *Gentlemen of the House of Commons* that Lord Russell "resumed his seat amid cheers," adding that some MPs could be heard whispering to each other Talleyrand's recent *mot*, "the Reform Parliament was the convocation of the Estates General, which at Paris had preceded the French Revolution."[26]

On March 2, the day after Lord Russell had introduced the bill, Thomas Macaulay, MP for Calne—a seat "in the pocket" of Lord Lansdowne—rose to encourage its passage. It was by all accounts an electrifying speech, and one that carries us back to Charles James Fox and his fear that if we did not reform Parliament, we would nurture radicalism. Here is part of Macaulay's closing:

> Save property divided against itself. Save the multitude, endangered by its own ungovernable passion. Save the aristocracy, endangered by its own unpopular power. Save the greatest and fairest, and most highly civilized community that ever existed, from calamities which may in a few days sweep away all the rich heritage of so many ages of wisdom and glory.[27]

Three weeks later the bill passed its second reading with a majority of one in what was the largest vote in parliamentary history, with 608 members in attendance. The elation, however, was short-lived as the bill suffered in subsequent committees. The Whigs decided to dissolve their government and seek a greater majority in new elections.

The new parliament met in June of 1831with a solid majority of over a hundred for reform. The second reform bill cleared the House of Commons easily, only to falter in the House of Lords by forty-one votes. It had been hoped that the Lords would defer to popular sentiment, but it was not to be.

Place realized reform was now caught between two dangers. On the one hand, the Whigs might capitulate in order to secure passage of some bill. On the other hand, increased agitation—there were riots in Derby, Nottingham, and Bristol—might lead to armed rebellion. Place opted to support a mass demonstration, a procession in London to present the King with a clear call for reform. It was successful, but rumors of compromise continued to float about.

The pressure for reform from both the middle class and the working class was relentless and in December the Whigs put forward the bill again. It passed the House of Commons with an even greater margin.

1832

This bill was sent to the House of Lords late in March of 1832. After several weeks of parliamentary maneuvers, it became clear that the Lords would not pass a bill comparable to that passed by the Commons. Lord Grey resigned and the Tories were given the task of forming a new government. The Duke of Wellington stepped forward and over the next several days tried to form a government.

Such are the practices of parliamentary democracy, but they only added fuel to the flame. As word of these events spread, the intensity of agitation ratcheted even higher. Place records that on Friday, May 11 he met with a delegation from Birmingham, reporting that they'd had a spontaneous gathering of a hundred thousand persons and had determined both not to pay taxes and to arm themselves. While these two options may seem incongruous—on the one hand civil disobedience, on the other armed rebellion—what they really signaled was the determination to do something.

The next day at another meeting at Place's home, it was decided that should the Duke succeed in forming an administration it was likely he would act at once to put down the people by force. It was also felt his administration would immediately provoke a general panic. Fearing such a panic would spur an armed rebellion, Place went back to the notion of withholding taxes. An economic strike of some sort might well be effective. We can see where this comes from in a comment Place made several years later as he reflected upon the failure of the Chartist Movement:

> All these persons thought as most of the politically associated working men still do, that—noise and clamour, threats, menaces and denunciations will operate upon the government, so as to produce fear in sufficient quantity to insure the adoption of the Charter—they have yet to learn that these notions and proceedings contain not one element of power—that the Government as mere matter of course will, as every Government must, hold people very cheap who mistake such

matters, as have been mentioned, for power . . . they have not a glimpse of their own, much less of the actual condition or relation of the several portions of society, who must concur, before any great organic change can be even put in progress.[28]

It was not that Place was a pacifist. He felt the threat of force among the working and underclasses must be accompanied by a clear signal that the middle classes are fellow travelers, equally discontented. Otherwise, they were simply playing onto the hands of those who wished to maintain the existing order. The Iron Duke, the victor at Waterloo, would not hesitate to call out the army.

The special thrust of an economic challenge to the government was very appealing, but Place was not keen on withholding taxes. He sought something bolder: a run on the banks. He took a large sheet of paper and wrote out this slogan: "To Stop the Duke, go for gold."[29]

England's might lay in its accumulated wealth. The empire was essentially a financial engine. The colonies were there to serve the needs of London's financial institutions. But this mechanism was fragile; it depended on confidence. Destroy the reputation of the banks and it would all crumble.

This was Place's idea. It would not be the first time anyone deliberately set out to drain a bank's reserves. In *Old and New London*, Walter Thornbury tells of a time early in the eighteenth century when the Bank of England set out to break the Child and Co. Bank, one of the oldest financial institutions in England, having been established in the 1670s. It was the practice in the early 1700s to give a receipt for the deposit of funds which could then be used in exchange for money. It was somewhat like a check. The Bank of England decided to collect such receipts and when they grew to a sizeable amount to go and demand the money, hoping the bank would not have sufficient funds to cover. The bank caught wind of the scheme and applied to an aristocratic friend, the Duchess of Marlborough, who gave them a check of £700,000 drawn on the Bank of England. Thus, when a clerk from the Bank of England arrived with a bag of receipts, Child's sent one of its own clerks to the Bank of England to cash the Duchess's check. By moving slowly through the receipts, Child's was able to cover the demand with the Bank of England's own money.[30]

The important difference with Place's scheme is that its motives were political, not financial. He didn't want the Bank of England to fail. He wanted to so scare the Duke so that he would capitulate and allow the passage of the Reform Bill. Place knew those who supported reform went far beyond the discontented working classes. The Industrial Revolution had created great wealth, both industrialists and a rising bourgeoisie. This new wealth was as disenfranchised as any other resident of London, or Manchester, Birmingham, or Leeds. Going for gold would make clear the power of this wealth.

If it worked. Grote was not convinced it would. A run on the banks is so unsettling; he feared the commercial interests would turn against reform. Place mulled over his friend's fears, but decided to stay the course.

Place's slogan energized the reformers. Money was put on the table and within hours bills were being printed and posted throughout London and in provincial centers. One and a half million pounds were withdrawn in just a few days. And heady days these were. In short order the House of Lords rejected reform, the Whigs dissolved their government, and the Tories could not form one of their own. It is often suggested that the Tories ultimately gave way because the King had let it be known that he would create sufficient new Whig peers to carry a vote through the House of Lords.[31] This may have been the case; it is hard to tell with so many rumors and speculations floating about. We do know that for over a week neither the Whigs nor Tories could navigate these turbulent waters. We also know the King was not happy about stacking the Lords. He demurred, as if such a ploy were too tawdry.

But though one might pause, hoping for some gambit that would unlock the political stalemate, the threat of economic turmoil was galvanizing.

The call for gold began on Monday, May 14. On Friday, May 18 a representative of the Bank of England met with Grey, Wellington, and the King, telling them of his fears of an imminent collapse of the bank. The Duke withdrew from his efforts to form a government. Earl Grey formed a new government, the run on the bank ceased, and the Tory peers of the House of Lords agreed to abstain rather than oppose reform. On June 2 the Reform Bill became law.[32]

But even before the bill had passed, the government made clear its concerns. On May 21, the first day of the new Whig government and only three days after the King had met with the representative of the Bank of England, a secret Parliamentary committee was set up to examine the condition of the Bank of England and whether changes should be initiated. The government was riveted by the question: was the Bank safe?

On July 20 Grote was a witness before this committee. When asked if he had any suggestions, he replied that the Bank should regularly report on its holdings. That is, it should make public how it stands during a run against its holdings. After some discussion of the nature of the reporting Grote thought advisable, Lord Althorp, a leading Whig minister and head of the committee, then asked the crucial question: "You are probably aware that in May last there was a considerable draft upon the treasure of the Bank of England, and the consequence of that demand was that the cash of the Bank of England became unusually low; do you think if the amount at the time had been published (that demand arising from political causes) the publication would not have created a material inducement to withdraw the whole of the treasure of the Bank of England that remained?"[33]

What an extraordinary phrase, "You are probably aware." Grote was widely known as a Benthamian reformer and had been right in the thick of things; though he had not been convinced that going for gold would work. Furthermore, on the day following the posting of the placards across London and key provincial cities, the *Evening Standard* accused Grote of having originated them. What an extraordinary situation. About a dozen witnesses appeared before the committee, including the governor of the Bank of England and two directors. Grote was the only Benthamian in the crowd.

Grote, by the way, replied "no" to Lord Althorp's question; that in a matter of political motives the publication of the state of the Bank's bullion holdings would not matter.

So?

The Reform Bill passed. Seats in the House of Commons were redistricted. Why is this significant? After all the fuss and bother, why would getting rid of a bunch of small boroughs matter so much? What sort of changes follow from redistricting?

Certainly trading Old Sarum, Dunwich, and a host of small boroughs for the likes of Leeds and Manchester was a move toward greater popular representation, but did this represent anything deeper? One might assume, for example, that the new electorate, coming as it did from more urban quarters, would be more committed to issues arising from industrialization and that there might be parliamentary commissions and subsequent legislation to limit abuses, and this was the case. More subtly and perhaps more importantly, the election for the new reformed Parliament attracted a host of candidates who had little to do with the aristocracy. More than sixty seats were taken from the pockets of the landed elite and turned over to an electorate less firmly tied to the lord of the manor.

There was a new game in town and people were excited. But the changes set into motion by the Reform Bill were deeper than parliamentary commissions or legislation.

If we go back to Mill's "Spirit of the Age," to his call for a social and moral revolution that would deny unearned distinctions and importance, we can feel that something more was in the air. The elite felt the challenge. Writing toward the end of the century, Thomas Escott goes so far as to suggest that herein lay the central push for one of the more profound changes in English society. The patrician elite redefined themselves, moving beyond such questions as "Who is he?" and "How much does he have a year?" to "What has he done?" Prestige by achievement replaced prestige by position.[34] Escott may well have caught that such a change had taken place by his own day, a new seriousness of purpose among the landed elite, but it is difficult to locate the moment such changes are set in motion. Doors are hard to find in clouds, especially in storm clouds.

The early nineteenth century was a period of marked change. Industrialization disrupted the traditional economy around the lord's estate and fed the rise of factory towns. We can see this in population numbers, in literature, in parliamentary commissions on the conditions of the working classes, and in the outbreak of urban plagues. We can also see it in the realization of a potential for a new kind of political protest: the mass march, now familiar, but then new. It is oftentimes hard to gauge the onset of the modern era, especially as so often change is a mixed bag. But the peaceful intimidation by a hundred thousand souls marching to underline their demand for Reform is certainly progress over a call to arms. This was a positive step in the practice of politics in a mass democracy.

At the same time, we need to recall that 1832 was a long time ago. In a charming book, *Age of Scandal*, T. H. White suggests that in the general sorting of ages and epochs in England we have overlooked a distinct era. Rather than see the march of sensibilities and events as stepping from the Enlightenment of the eighteenth century to the Industrial Age of the nineteenth, he suggests we interpose England under the Georges—roughly the stretch from the 1740s to the 1820s—as its own entity. Hume and Voltaire notwithstanding, England at this time was still a very Christian nation with a rather small landed elite of fewer than two hundred titled families who still ran things very much as they had over the centuries. Quoting G. M. Trevelyan, he observes that "it was an age of aristocracy and liberty; of the rule of law and the absence of reform."[35]

It was also, White tells us, a time which valued "bottom." Life was hard, and it was crucial to be prepared to handle its pains and sufferings. "Bottom" was a nautical term, in praise of the capacity to hold one's course, despite the wash of tempest and storm. And here we see most clearly how distant this world was at bottom. It was an age of curious elixirs: turnip water, snail tea, or a posset, a drink made with warmed ale or wine and milk, which became medicinal by adding horse dung.[36] We forget how necessary stoic acceptance would have been in an age before anesthetics and modern medicine. Childbirth, for example, was often deadly.

Eighteen thirty-two was a long time ago, but it was also charged with elements of the modern era. The Reform Bill would seem to mark that moment when people collectively acknowledged a tumble of transformations. They no longer lived in their parents' England. Recall Mill's conviction that they had outgrown old institutions and doctrines, and further that old bonds no longer held. The Reform Bill marked the public affirmation that this was indeed so. In this sense, we can agree with Escott when he links an act of parliamentary redistricting to pronounced changes in class norms. Reform was a statement about old and new ways.

"Old and new ways" seems just the right note to sound as we move from the marching, charging feet of tens of thousands of protesters first to the quiet study of the classical scholar and then to the fieldwork of the geologist, before we settle on a young man's encounter with the cruelty of life in Tierra del Fuego.

And so we turn to the Athenaeum and a meeting of Connop Thirlwall and George Grote.

NOTES

1. Gregory, *Second Cuckoo*, 154–56. We may add that Charles Lyell was proposed to be the MP for the University of London in 1861, but he declined; see Lyell, *Life, Letters*, vol. ii, 343. This notion of representing leading interests continues to this day in Hong Kong, where half the seats in the city council are "functional constituencies" linked to industries rather than population. See Hilgers, "The Rise of Hong Kong's Democracy Movement."
2. Trevelyan, *Illustrated English Social History*, 109.
3. Fox, *Speeches*, 714.
4. *Ibid.*, 688.
5. *Ibid.*, 676.
6. Smith, *Wealth of Nations*, vol. i, 6–7. It is worth noting that at this time it was still the case that most homes were workplaces of one sort or another. Only later with the development of factories would the home become essentially a residence; see Jackson, *The Necessity for Ruins*, especially 116–19.
7. Schweber, "Scientists as Intellectuals," 4.

8. Kay, *Moral and Physical Condition*. Kay's study, we may add, was cited by Friedrich Engels in his *Condition of the Working Class in England in 1844*.

9. *Ibid.*, 25, 13.

10. Mill, "The Spirit of the Age," *Collected Works*, 228–29, 230, 245. There were four instalments of this essay in the *Examiner*, all of them collected in vol. xxii of the *Collected Works*. See also Himmelfarb, *Spirit of the Age*.

11. Grote, "The Essentials of Parliamentary Reform," in Bain, *Minor Works*, 17.

12. *Ibid.*, 12–13.

13. *Ibid.*, 11.

14. *Ibid.*, 15.

15. *Ibid.*

16. Wallas, *Life of Francis Place*, 13–14.

17. Smith, "Bentham on Fallacies," 209.

18. Wallas, *Life of Francis Place*, 186–89.

19. *Ibid.*, 206.

20. *Ibid.*, 141.

21. *Ibid.*, 144.

22. *Ibid.*, 145–46.

23. *Ibid.*, 244.

24. *Ibid.*, 250.

25. Clark, *Old Friends*, 108–09; Wallas, *Life of Francis Place*, 256.

26. Escott, *Gentlemen*, vol. ii, 274.

27. Macaulay, "Parliamentary address of March 2, 1831," 162.

28. Place, "Letter to Harrison," in Rowe, *London Radicalism*, vol. v, 228.

29. Wallas, *Life of Francis Place*, 310.

30. Thornbury, *Old and New London*, 461.

31. For example, in Fraser's *Perilous Question*.

32. Wallas, *Life of Francis Place*, 320–21.

33. *Report from the Committee of Secrecy*, 368.

34. Escott, *England*, 318–19. There is an echo of Escott's thesis in a story Jerome Bruner tells about the historian Sir Alan Bullock, who found himself seated alongside Queen Elizabeth. Knowing of her disdain for small talk, Sir Alan asked her when the Royal Family had decided to become respectable. She replied, "it was during Victoria's reign, when it was realized that the middle class had become central to Britain's prosperity and stability." Bruner, *Culture of Education*, 128.

35. White, *Age of Scandal*, 30.

36. *Ibid.*, 38–40.

3

A MEETING AT THE ATHENAEUM

Imagine two young men, old friends, have arranged to meet at the Athenaeum, a London club. Such clubs already had a long history by 1832, providing a home away from home for the landed elite who would come up to London for the season. The Athenaeum was somewhat different, founded only a few years earlier for individuals distinguished in science, literature, or the arts, as well as those who were patrons of such efforts: a meeting place for the mind. As such it signals the rise of the public intellectual and the rise of a London season for scholars, most notably scientists and their many new societies, including the Geological Society, whose annual meeting was the prompt that brought Thirlwall to London this February.

Both Grote and Thirlwall have gained their stride since their school days together, accomplished significant things with the promise of more to come. Connop Thirlwall is thirty-four in February 1832, and George Grote is thirty-seven.

Thirlwall had been a most precocious child; he had learned to read so well at so young an age that he was taught Latin at three and Greek at four. It was said that at Charterhouse, an old red brick public school that had been founded in the early seventeenth century, the young Thirlwall did not care to join in on the games of the other boys, but would withdraw to some corner with a pile of books. The child was father to the man, and all his life he carried books with him wherever he went. But more than this, Thirlwall retained a certain freshness and gentleness across his many years, and with it the capacity to enjoy life's more simple pleasures.

Ironically, a collection of his letters begins with one written when he was twelve, in which he confidently affirms both that Oxford was much the better university and that classical studies are no longer relevant to men of affairs. Despite such sentiments, Thirlwall would leave Charterhouse for Trinity College, Cambridge and a career much indebted to classical studies. Thirlwall sparkled at Trinity, earning a fellowship which enabled him to travel. He spent just over a year on the continent, living "the most enchanting of my day-dreams." When he returned he began study at Lincoln's Inn Fields, hoping that law would enable him to pass "a quiet but not indolent life, obscure but not useless," and looking forward "to contemplating the sights and sounds of nature and the finest productions of the human intellect."[1]

There was much to occupy Thirlwall in London in mid-1820. John Stuart Mill tells of hearing Thirlwall in a debate at the Co-operative Society, finding him to be much the best debater he had heard till then and indeed ever. These debates proved to be so stimulating that a small coterie formed a debating society of their own, including not only

Thirlwall and Mill, but also Macaulay, Edward Bulwer-Lytton, Earl Grey, and another prominent Whig, Samuel Romilly. Sadly, they could not maintain the intensity and the society faded, as did Thirlwall's patience with the law.[2] Shortly after entering the Bar, he chose to return to Cambridge to finish out his fellowship.

Before this, midst the various demands of the law and the pleasures of debate, Thirlwall had thrown himself into another project. He translated a piece by a German theologian and historian, Friedrich Schleiermacher's *Essay on the Gospel According to Luke*, penning as well a lengthy preface on the nature of inspiration. This had been truly engaging, and it became clear to him that this was his calling.

He left law school and returned to Cambridge as a fellow, continuing his effort to bring recent German historical scholarship to an English audience by translating with Julius Hare, another Charterhouse friend and now a colleague at Trinity College, Niebuhr's *History of Rome*. Both works would roil the relative calm of English waters. By 1832 Thirlwall and Hare, having completed the three volumes of Niebuhr's *History*, were editing a new journal of classical studies, *The Philological Museum*, and soon Thirlwall would begin writing his *History of Greece*.

Thirlwall's stay at Cambridge as a fellow would be marked by earnest enthusiasm as a scholar, as a teacher, and as one of a small group dedicated to university and political reform. But it was an unexpectedly short stay. It was abruptly brought to an end in 1834 because of a dispute over Chapel attendance, an aspect of the larger issue of the university's ties to the Church of England and her refusal to admit dissenters—those of a Christian faith but not the Church of England. The matter had come before Parliament, and some in the university defended the practice of requiring an oath of allegiance. Thirlwall replied that Cambridge was not a divinity school and while they were at it, they should take this opportunity to cease requiring daily chapel attendance. The cause of religion, he felt, was best served on a personal, rather than institutional basis. The Master of Trinity heartily disagreed, writing that the alternative was between a "compulsory religion and no religion at all." To which Thirlwall replied that "the difference between a compulsory religion and no religion at all was too subtle for his grasp." The Master asked him to step down, which he did.[3]

John Stuart Mill comments on this event, calling Thirlwall the "first scholar in Great Britain and the only clergyman of the Church of England who has acquired a European reputation," adding that he had been "ejected from his lectureship in the most liberal college of the two universities" for asserting what everyone knows to be true—that the university did not give religious instruction.[4]

Once again Thirlwall packed his bags, this time to become a parish priest in Kirby Underdale, Yorkshire, where he served a small rural community of perhaps three hundred. He had already completed the first volume of his *History of Greece*, and now proceeded to throw himself into the project, working at times sixteen hours a day, finishing the remaining seven volumes by 1840. Such discipline may often be linked to a driving ambition, but no doubt the real push here was a reaction to having been asked to leave Trinity. He would not allow the narrow thinking of those like Christopher Wordsworth, the Master of Trinity, to stop him. He no longer had access to the rich holdings of the Wren Library at Trinity College, but he would make do, and he did.

In fact, he did very well. The *History* was a great success, not least because it led to his appointment as Bishop of St. David's where Thirlwall came down squarely where he was meant to be.

Actually he came down in two places. The cathedral of St. David's is in the town of St. David's, in westernmost Wales, on a rugged promontory jutting into the Irish Sea. Here's a description from the *Penny Cyclopaedia* of 1837:

> It was antiently large and populous, and during the middle ages was the resort of a great number of pilgrims. At present its appearance is that of a poor village, the houses, excepting those of the clergy, being mostly in a ruinous state. The locality is lonely, and the neighbouring district wild and unimproved . . . In 1831 the population of the parish . . . was 2388. Druidical remains are numerous in the neighbourhood, consisting of sepulchral heaps of stones, barrows, tumuli, holy wells, and some antient fortifications.[5]

Quite the setting, like something out of a gothic romance, or perhaps like Dylan Thomas's description of a village by the sea:

> Quite early one morning in the winter in Wales, by the sea that was lying down still and green as grass after a night of tar-black howling and rolling, I went out of the house . . . to see if it was raining still, if the outhouse had been blown away, potatoes, shears, rat-killer, shrimp-nets, and tins of rusty nails aloft on the wind, and if all the cliffs were left.
>
> The town was not yet awake. The milkman lay still lost in the clangour and music of his Welsh-spoken dreams, the wish-fulfilled tenor voices more powerful than Caruso's . . . Miscellaneous retired sea captains emerged for a second from deeper waves than ever tossed their boats, then drowned again, going down down into a perhaps Mediterranean-blue cabin of sleep . . . Landladies, shawled and bloused and aproned with sleep in the curtained, bombazine-black of their once spare rooms, remembered their loves, their bills, their visitors, dead, decamped, or buried in English deserts.[6]

The cathedral stands on the grounds of a monastery established late in the life of St. David, the patron saint of Wales, tracing itself back to the year six hundred and one. It had a long and troubled career, having been sacked twice by Vikings and suffered many another hardship. The original Bishop's Palace that had stood alongside the cathedral was in ruins, and as long ago as the sixteenth century a new residence had been established in Abergwili, some fifty miles away in the gentle folds of the countryside outside Carmarthen. Both were a long way from London, from its bustle and busy-ness, from its thirst for growth and gain, but Thirlwall loved this countryside and he would often walk about.

He added Welsh to his many languages, surprising all by preaching a sermon in this ancient Celtic tongue within six months of his appointment, and he took the administration of the See most seriously, as we can see in this snippet from a biographical sketch by John Morgan, the Archdeacon of Bangor:

> The unwearied energy, the unsleeping vigilance, the high and unsullied character, the transcendent talents, the wide and commanding influence, were all his own, the essential and inseparable qualities of the man, and these he devoted with all his heart to the advancement of the diocese. The fruits were to be seen in the new schools that met us in every hamlet, in the training college at Carmarthen, which his lordship opened in 1848, in the large number of new churches or churches restored, in the increased activity of the clergy, and in the reviving attachment of the people to their ancestral Church.[7]

To this we may add the more personal observation of Arthur Stanley, Dean of Westminster, from his eulogy at Thirlwall's funeral. He sought to capture the essential and inseparable qualities of his friend, speaking of his judgment. Thirlwall had been, he said, "judicious . . . exactly that quality of judgement, discretion, discrimination which is the chief characteristic of the biblical virtue of wisdom."[8] It was a quality at the very soul of the man.

As a young scholar Thirlwall's translations of the work of Schleiermacher and Niebuhr had disturbed the relative calm of the Church. But now, as Bishop, his learning and judicious evaluations would calm the deep waves that broke against the Church, offering a rock to stand upon.

In one of Plato's dialogues, the *Phaedrus*, Socrates expresses a fear that writing would lead people to reflect less.[9] Such a fear would have been wasted on Thirlwall. He read not for the distraction but for the invitation. He loved the many-sidedness of things. You might think you are looking at a cube, but with closer examination it turns out each corner has been clipped and crisp points have been replaced by tiny flat surfaces. The clarity of six clean sides, each a neat square, has given way to a fourteen-sided amalgam of triangles and octagons. Thirlwall loved the "fourteen-sidedness" of things.

In an article that would not appear until after he had died, Thirlwall discussed the Broad Church school and it strikes just this note of the fourteen-sidedness of things, but first a word on the notion of a "broad" church. There has long been a distinction between "high" and "low" church, where "high" favored ritual and formal trappings and "low" replaced formality with enthusiasm, an enthusiasm often dressed in plain white. The "broad" churchman, however, was neither high nor low, standing instead in opposition to the "narrow" churchman and favoring a national church which embraced a wide field of sensibilities.

It seems a Dr. Littledale had written that the faults of the Broad Church school were due to ignorance and that "theology grows clearer with the advance of knowledge." To which Thirlwall replied that "it would hardly be possible to frame a proposition more entirely contrary to all the results of my study of Ecclesiastical Theology, or to those of my personal experience." In theology as in life, the more we know, the more complicated things become—which was exactly what Thirlwall loved about learning.[10]

Thirlwall would be Bishop of St. David's for over thirty-four years, a strong leader both within his See and for the nation as a member of the House of Lords. Though he had bounced around a bit from Cambridge to the continent, to Lincoln's Inn back to Cambridge and on to Kirby Underdale, he had only felt out of place for that brief stint as a lawyer. A long life of influential acts and words.

Now a word about George Grote, who had also attended Charterhouse, leaving in 1810 to work at the family business, a bank. Instead of the elegant buildings and quiet quadrangles of an ancient university, the young Grote made his way across muddy London streets busy with commerce.

Handsome, with an athletic presence, Grote led a busy life: not so much like Sandburg's ragtime jig whistled down toward the sunset, as like a piece by Beethoven, both passionate and rational. By 1832 Grote, too, had penned works that had somewhat roiled English waters. As we saw in the chapter on the Reform Bill, Grote had just composed an influential pamphlet on Parliamentary reform. He had also earlier edited a batch of Jeremy Bentham's manuscripts on matters concerning natural religion, yielding *An Analysis of the Influence of Natural Religion On the Temporal Happiness of Mankind*. Back in 1826 Grote had made his first foray into classical studies with an essay on Mitford's *History of Greece* for the *Westminster Review*. In 1826 he also threw himself into work for the newly established London University, where he helped design its faculties and curricula. Grote would soon become a Member of Parliament for the City in the new reformed Parliament. Still later he would serve on the board of the British Museum and become President of London University—quite remarkable for a man whose schooling had ended at the age of fifteen. With all this, he is remembered chiefly for his twelve-volume *History of Greece*, a work so commanding that it framed scholarship for generations to come.[11]

Again, a long life of influential acts and words.

Let us now conjure up an evening in February 1832. They meet for the first time in a while and make their way to the drawing room where they might sit and chat. The room ran the width of the building, with a tall ceiling, sofas, the occasional table and chairs, bookcases and cabinets with busts of scholars, artists, and civic leaders. It was a comfortable place for a quiet read or conversation. After catching up on how things had been since last they'd met and how John, George's younger brother, was faring at Trinity, conversation quickly turns to work, and the value of colleagues. Thirlwall is talking . . .

"I'm up for the annual meeting of the Geological Society tomorrow evening. There'll be dinner at the Crown and Anchor; perfect for the cold and dank weather we've been having with its cozy atmosphere within despite its shabby exterior. Then we'll move on to the society's chambers at Somerset House for the presidential address. Sedgwick has been a delight these last two years. Gracious, yet sharp. No empty platitudes. He took advantage of the platform to talk about issues and to explain the best thinking. It will be Murchison in the chair this time. Do you know him? A solid geologist no doubt, but not, I fear, up to Sedgwick's mark.

"Sedgwick and Whewell, along with Hare of course, are the heart of what has made Trinity so vital for me. Rather literally heart-like, come to think of it: two parts

to the heart-beat. Whewell is that energetic first push to get things going. I know he often seems rather stiff and austere, but really he's a Lancaster lad, full of bounce and good spirit, ready to take on all comers. There's a difference with Sedgwick. He's every bit as energetic, but it carries a different way. He seems rough and tumble, but actually he's more reserved and reflective. They're natural complements.

"Adam's not that much older than the rest of us, but there's a quality to him that makes him father to us all . . . a manly solidity, as if he had planted his feet, faced the universe, and had come away with lesson learned. I know you know him, but one can get the wrong impression . . . he can seem rather last century at times, but he's not."

"Oh, I have no problem with your Adam, Con; nor Whewell for that matter," Grote replied. "Sedgwick's engaging and earnest, and master of both geology and a good story. You clearly have found a comfortable home at Trinity. Good talk is a rare commodity. I find it easy to envy you your quiet quads and walks along the Cam. It is all so delightfully far removed from the constant bother of life in the city."

Grote paused, turning away for a moment. "I hated not being able to go down to university, but father would have nothing to do with it. I would get up every day at four in the morning to do my own studies before I went to the bank, but that's a poor shadow of what I imagine university life was like. My work with the new London University over the last half-dozen years has really been me trying to make up for that lost chance. And now that is done. I had to step away from the council; they were going to weaken the very heart of the project, and I couldn't persuade them otherwise."

"It's I that envies you, George; though I can appreciate what you are talking about. Back when I first went down to Cambridge . . . I don't know if you know this story, I was met at the coach by a porter who loaded my box onto his trolley. As we walked through town, over the cobbled streets, I kept looking for signs of the university. All I saw were shops, book stalls, butchers and bakers, and a great warehouse on our left with a plain brick façade. Then the porter turned sharply down a narrow passageway, into the warehouse! I wondered what he was up to, but as we passed through that tunnel into the courtyard within I gasped at its beauty. The Trinity quad is a marvel.

"At the same time, I also felt myself stepping out of that mix of things which was so much a part of life when we were going to school in London. The university had literally turned its back on the town. I felt I had come to a private world. We would labor in fields of ancient tomes and commentary, of postulates and scholia, free of the mud and misery of everyday life in town.

"I think it's a mistake. So does Coleridge. Have you read Church and State? There's so much in it about what learning and culture are really about . . . In any case, I think Cambridge is not much longer for me. Hare is leaving and it's hard for me to imagine staying very long with him gone."

"That's the thing, Con. I am convinced you and Hare and your friends at Trinity are waging a losing battle. Despite the thoughtfulness of so many over the centuries, the church routinely turns it back on its flock. It's in its nature. But then again, it is no doubt a battle you cannot turn your back upon. It is in your nature.

"I am sorry Hare will be leaving you. Listening to you talk about your club of fellows I can't help but be reminded of how elated I was when I first met Ricardo . . . goodness, that must be almost fifteen years ago now. He introduced me to Mill and before I knew it I was spending hours at Bentham's, soaking up everything from politics and history to law and theology. Any number of folk will gather, but the core has been Bentham, the two Mills, and Francis Place. I'm not sure you've met him. What an extraordinary character. Been through fires you and I could hardly imagine. You might think it would have hardened him, but quite the opposite. He's a most sympathetic being, always working so that others may lead lives where good work is rewarded.

"You know that Harriet and I lost our only child. Place had also lost a child, and he's been marvelous for Harriet; writes her all the time."

After a pause, Thirlwall remarks that he had read Grote's pamphlet on reform and found it very persuasive, not that he needed any convincing . . . but still. "I especially liked the way you moved beyond pocket boroughs to the deeper matter of expanding the electorate and fostering a more inclusive democracy. Do you see the common man coming to a fuller awakening?"

"You've always been such an optimist, Con. I see it more as a move toward balancing the complete control of things by the elite with working men able to assert their self-interests more effectively and so not having to resort to the threat of strikes and violence. But, of course, it would be grand if they were to rise to the occasion and allow public matters to be complicated and worthy of study and expertise and so elevate public discourse. It would be a privilege to serve in Parliament if such were the case.

"Since writing the pamphlet matters have become so intense, so charged it reminds me of a trip Harriet and I took to France in the Spring of 1830 . . . on the eve of the revolution. What a trip that was. In a way it began for us in 1824 or so. That's when Charles Comte was here, exiled by the now deposed Charles X. Did you meet him?"

"No, I don't know that we met personally, but I remember him. That was when I was at Lincoln's Inn and happy for any distraction."

"Well, he's a fine fellow. Married Say's daughter . . . Say, an economist and a good liberal, had been active in the Revolution, uncomfortable with Napoleon and more so under Charles. Anyway, in time Comte returned to France and Harriet stayed in touch with both him and his father-in-law. We stayed with them on our visit and they took great care of us. They even arranged for us to stay with Lafayette at his estate. Just as we arrived, the despot Charles had dissolved the Chamber of Deputies. Everyone was agitated. Things were coming to a head. Lafayette was the leader of the liberals in the Chamber and everyone turned to him for what to do next.

"What a figure! Off to America as a young man, he becomes aide de camp to Washington. He returns to France, but his idealism had been set in stone by the American struggle and he is unable to abide the 'Ancien Regime' and its many abuses. It all tumbled so quickly . . . his clothes might still have been damp from crossing the Atlantic as he led the storming of the Bastille.

"Reflecting on these events one evening, he told us how he had been given the 'lock-stone' of the Bastille, the large granite block that held the bolt mechanism of that dreadful prison. It was his greatest treasure, but he said he could not keep it. He sent it to the man who really deserved it, George Washington . . . Father of two revolutions," he said.

"And now, at seventy-three, Lafayette was at the center of a third revolution. But we had to leave. We got word that my father, already very ill, had turned for the worse. He had never agreed with my politics any more than my desire to go to university. His final gesture would once again take me away from something I truly valued. We later learned the committee that negotiated with Charles over his departure had approached Lafayette, asking him to take over the government, but he turned them down. It was Lafayette who said France should be led by Louis Philippe. What a grand figure!"[12]

Then Grote went on to talk about the press for parliamentary reform, recent events and what might happen next.

George Grote and Connop Thirlwall were scholars of the first order whose contributions were of immediate and long-lasting consequence. They stepped forward from rich intellectual settings and the histories of antiquity they wrote were as rich and as different as those settings. In Grote and Thirlwall we shall find two coherent expressions of the character of history, perspectives intimate to the deep currents of their day, currents whose head waters stemmed from Bentham and Coleridge.

HISTORY AS CONCEPTUAL DEVICE

As we begin to examine classical scholarship early in the nineteenth century, we should note that Greek and Latin had long been at the very center of learning. They were the languages of the New Testament, as well as Homer, Virgil, Aristotle, and Livy. And in publications like the *Edinburgh* and *Quarterly* reviews, it was not uncommon to encounter quotations in Greek and Latin without translation, even unattributed; simply introduced with a casual "as the poet said." The likely reader would know well both Greek and Latin.

In his history of classical scholarship in England, M. L. Clarke takes Thirlwall's lectures across his brief tenure at Trinity as the beginning of a new era. These lectures encompassed far more than the study of the Greek language which had long been the standard practice. As Thirlwall explained, the study of ancient philology had addressed "an infinitely minute branch of the subject and has been severed from the rest and treated as the whole." Thirlwall replaced the narrow, exacting analysis of Greek associated with the great seventeenth-century Trinity scholar, Richard Bentley, with the more expansive German historical approach.[13]

A more immediate judgment of Thirlwall's impact comes from Adam Sedgwick. In December of 1832, Sedgwick delivered the sermon at an annual commemoration service at Trinity College. He took advantage of the occasion to re-examine the meaning of a Christian education in this new epoch. His talk was so compelling he was prevailed upon to have it published, and so we have *A Discourse on the Studies of the University.*

In reviewing the recent accomplishments of Trinity's scholars, Sedgwick praised the way classical studies was being taught. Acknowledging the master work of Bentley, Sedgwick nevertheless frowned upon its overemphasis on style. Substance had been overlooked. The philosophy and ethical writings of the ancients had been ignored. We have been too taken, he continued, with "the measure, the garb, and fashion of ancient song, without looking to its living soul or feeling its inspiration."[14]

We may complement this hasty sprint across Thirlwall's tenure at Trinity with a few fragments that speak more to the spirit of the place. Take this delightful episode from 1829, when a petition was presented to the University Senate protesting the granting of relief to Catholics. The matter of Catholic relief involved granting to Catholics civic rights and a fuller place in the English polity. Both Cambridge and Oxford were institutions of the Church of England. While those of other faiths—Catholics, Dissenters and Jews—could take courses at the university, one could not get a degree without subscribing to the articles of the Anglican faith. The more liberal members of the Senate, chief among their small number Sedgwick, Thirlwall, and William Whewell, prolonged the proceedings on this particular occasion, demanding that a roll call be taken. The result of the final vote was to deny the petition and so help support relief, a matter quite surprising but explained for us in an entry in the *Cambridge Chronicle*:

> The result appears to have been principally owing to the somewhat unexpected arrival of several members of the Inns of Court, who came down for the express purpose of voting upon the occasion; two Paddington coaches with full complements of inside and outside passengers arrived between one and two o'clock, returned to London the same afternoon.

The *Chronicle* did not approve of the unexpected arrival of the Paddington cohort. Their arrival, however, had not been unexpected. Our Trinity liberals had been in touch with a sympathetic friend and former fellow, the historian Thomas B. Macaulay, soon to be an MP, who had organized the day's jaunt from London.[15]

While Thirlwall and the Trinity cohort carried the day this once, reform at the universities was difficult. Yet they were steadfast in pursuit of a variety of reforms, several of which were successful. They played a central role, for example, in expanding the Tripos examination which was the final examination for graduation. It had been entirely given to mathematics; then in 1824 one could opt for a Classics Tripos. Our Trinity cohort worked to add a natural philosophy Tripos and then a moral philosophy Tripos.

Despite such efforts, our Trinity group is not accorded an honored place in the story of university reform. Leading critics, such as Lyell and Huxley, were outside of the Church, viewing Oxford and Cambridge as outmoded religious institutions afraid to embrace the modern world. They were especially critical of the rule which prevented resident fellows of the colleges from marrying. Our Trinity fellows agreed that the universities should be opened up and the cobwebs of an antiquated monastic heritage swept away. Yet there was much they cherished. They had seen it work. They had felt the way it fostered community and fellowship both rich and vital.

It was important to change where change was called for, but not at the cost of the very heart of things. For teachers to be resident and single made possible a heightened academic community and staved off the fragmentation which would naturally arise from each member of the community having their own household. We may catch a glimpse of the "table talk" of this cluster at Trinity in a comment from Hurrell Froude, an Oxford tutor and influential theologian. "Certainly these Cambridge men are wonderful fellows," he wrote. "They know everything, examine everything, and dogmatize about everything; they have paid particular attention to the geological structure of this place, and the botany of that, and the agriculture of another, and they are antiquaries, and artists, and scholars . . . Whewell's book, and Sedgwick's Lectures, and Thirlwall's research, and Hare's taste pop upon one at every turn."[16]

Even at our remove we can appreciate what Froude is talking about. The published works of this group has an extraordinary breadth. Sedgwick's *Discourse*, for example, surveyed the studies at the University and became with the greatly expanded fifth edition an encyclopedic commentary on everything from Locke to fossil fish, from chemistry to Bishop Butler's *Analogy*.

Books, lectures, artistic and literary taste, scholarly research: we can almost hear the conversations of the combination rooms at Trinity. Everything from the Cambrian system to the gods of Mt. Olympus, from Genesis to calculus, from Coleridge to Bentham, from facts to fiction, fables and fossils—all of these were on the table, to be examined and given place.

Let's try our hand at ancient myth.

GREEK HISTORY BEFORE THIRLWALL

We should begin with a modest step backward and consider William Mitford's *History of Greece*, the first volume of which appeared in 1784 and the last in 1810, twenty-six years later. Much of the attention Mitford's *History* drew derived from its outspoken politics. Mitford, a gentleman Tory, was sharply critical of Athens, the great democratic experiment. And certainly the times were right in this regard; his *History* was published across an era marked by fear of the mob—be it Thomas Paine's or Robespierre's.

Mitford's *History* was not dull; he was too alive to the political lessons at stake for that. As Thomas Arnold observed: "He described the popular party in Athens just as he would have described the Whigs of England; he was unjust to Demosthenes because he would have been unjust to Mr. Fox."[17] The immediacy of ancient Greece for Mitford was more than a matter of political parallels. He wrote confidently of an age that was not far distant in time or aspect. The passage of time had not clouded history's vision. He respected the authority of the ancients and drew directly from their "judicious" commentaries a rational account of key political and cultural events, even as he wrote about the earliest chapters of antiquity.

Here, for example, is Mitford on the founding of Sicyon, by tradition the oldest of the Greek city states: "It has been computed by chronologers, who have found credit with some of the most learned even of the present age, that Sicyon was founded two thousand and eighty-nine years before the Christian era and fifty-nine after the Flood."[18] Here is an historical time both definite and absolute, a framework from which events could be hung like ornaments on a Christmas tree, starting with Noah's flood, 2,148 years before the Christian era.

Migrations and colonial expansions were a central motif in Mitford's chronology. Alluding to *Genesis* and the Tower of Babel, Mitford spoke of the peculiar energy which had fired the souls of humankind in its infancy, as they migrated across the globe. Further, it was clear that civility and knowledge had only been preserved among a small number; while the rest had degenerated into savagery. This yielded a core theme of Mitford's study: a tracing of the re-emergence of civility. Within this trace, inventions in custom or technology were discrete events. Marriage had been brought to ancient Greece by Cecrops, political order by Minos. Danaus taught them to dig wells, and Cadmus taught them letters. And so, as Mitford put it, "the colonies from Egypt, Phenicia, and Thrace swiftly made the Atticans a new people."[19]

The key figure in Mitford's analysis was the poet, who wove together fact and fancy. The poet practiced a creative license, and so history became legend. The challenge before the historian was to reverse this process and regain history from the fabric of ancient song. Here is Mitford on the adventures of Jason and the Argonauts: "We do not believe all the romantic, and still less the impossible tales, which poets, and even some grave historians, have told of those famous adventures; . . . yet it seems unreasonable to discredit intirely the argonautic expedition."[20] But what credit could critical history give these legends? Mitford placed Jason and his crew within a coarse age that honored piracy. The aim of the voyage had been the Princess Medea and the riches of her kingdom. The magical fleece of gold was not, Mitford assures us, truly magical. It was a symbol of great wealth, derived from the fact that fleece had been used by the ancients to pan gold from streams, the tight fibers of the wool catching the flecks of gold as they passed by. And so the outrageous proportions of ancient legend were reduced to far more prosaic measure.

This glance at Mitford's *History* reveals several leading features. Historical agency was finite and discrete. There is no development to history, rather a series of particular items: adventures, colonies founded, inventions, and more. These were then arranged in chronological sequence, relying on the count of generations, a matter involving careful scrutiny. Finally, Mitford was a skeptic, a rationalist, reducing the proportions of ancient legend to more prosaic measure, all within a biblical time frame.

THE NEW MYTHOS

With this as our setting, let us turn to Thirlwall's work. We may begin with an intriguing footnote in a review of a recent book by Friedrich Kruse that Thirlwall wrote in 1832 for the *Philological Museum*, the journal he and Hare edited. In the body of the article Thirlwall refers to the "myths" of ancient Greece—as opposed, say, to legends—and the footnote is about his use of this word. He begins by acknowledging K. O. Muller's important discussion of myth as fable which was an important step in the right direction. However, not all myths should be seen as simply invented out of whole cloth;

some, those connected with historical events for example, should be acknowledged as having a more complex genesis. Such historical narratives may have sacrificed truth for the sake of a good tale; or they may have originated in the desire to make a confused tale more clear and distinct. "But many others again may have sprung up gradually and spontaneously, without any deliberate purpose or motive, and may be derived from the imagination not of one individual, freely exercising his inventive faculty for a certain end, but of a people, or a great number of individuals, who by a process, of which examples occur every day, may unconsciously and undesignedly modify a tradition founded on a real fact by successive additions and alterations till not a particle of it retains its genuine shape."[21]

This footnote is a slash of red paint. Like Turner, Thirlwall is following a new aesthetic. The canvas was not quite right. He needed to mark this spot and focus the reader's eye. Like Turner, the issue was authenticity and the atmosphere of a moment, but it was not about the "visual feel" so much as the "imaginative feel" of that moment, the spirit or genius inherent in ancient storytelling.

Going back to Plato's *Phaedrus*, Socrates talks about the difference between writing and living speech, offering an epistemological critique. Writing does not get at the truth the way discourse can. You cannot question the written word; it "goes on telling you the same thing for ever." But with the back and forth of a dialogue knowledge "can defend itself, and knows to whom it should speak and to whom it should say nothing."[22] This, however, is not quite the issue. Thirlwall is not interested in whether myths *were true*, but what it means that they *rang true*.

Myth was anchored in a creativity far different from storytelling as it is now commonly understood. Myths, like Socrates's "living speech," were oral "documents" and oral storytelling is a "be there." This difference is reflected in two words coined long ago. An early term for one who spoke was a "rhetor," where rhetoric is the art of persuasion. Speaking is interactive. You engage your audience, responding to their reactions. An oral document is, in this sense, distinctly interactive. This is in contrast to the word "author," which is associated with a set of words affirming the self, derived from the prefix "auth" and "auto," as in the word "authentic" where something is rightly linked to a particular agent and "automotive," a self-powered vehicle.

Myth as an oral document was interactive in its origins, in contrast to the self-directed author who can only guess at his or her audience's reaction. Because of this interactive quality, those ancient tales that thrived would have spoken to the sentiments and sensibilities of the community as a whole: it was the collective voice, as opposed to an individual's take on things.[23]

More particularly, the distinct feature of oral traditions for Thirlwall is that they are not fixed, remaining alive to adapt to new issues. How the past is understood is then drawn upon to meet the demands of the present, with "successive additions and alterations" which will carry a tale beyond its origins. The successive tales trace the changing sensibilities of a people.

To appreciate what's at issue in this slash of red paint, let's explore a central example: the exploits of Hercules. The tales surrounding Hercules ranged from the extraordinary—changing the course of a river to cleanse the Augean stables or wrestling with the god of death, Thanatos, to bring a noble woman back to life—to the more ordinary, such as punishing wrongs like robbery or sacrilege.

Mitford had approached these legends with a jaundiced eye, prying apart the weave of fact and fancy, scrubbing away the exaggerations of the ancient bards. He denied the labors of Hercules had any basis in historical events. They were mere poetic fancy. The real Hercules was a figure more like Sir Galahad, a knight errant wielding the sword of justice in an era when governments were too weak to meet this charge. There were for him two sorts of tale, an original kernel of actual knight-errant adventure and the fantastic labors of a hero fashioned by poets out of the fabric of their imaginations.[24]

Thirlwall's 1832 footnote is not the right place to illustrate what he was getting at, but if we turn to the first volume of his *History of Greece*, completed before he left Cambridge in 1834, we can see that slash of red paint turned into a buoy. In his analysis of the myths surrounding Hercules we find him distinguishing two sets of Herculean adventures, as had Mitford, but he takes the differences between cleansing the Augean stables and rescuing a damsel in distress as differences in the sensibilities and concerns of the audience, not the difference between poetic imagination and real events.

For Thirlwall, the interactive character of the creation of myths removes it from the common notion of storytelling—as hinging on the inventive faculty of the author— and shifts it to the collective mind. The labors of Hercules were early tales reflecting the wonder of folk within newly established communities: how did fields get cleared of beasts and boulders? Why does a river run past our settlement and not elsewhere? The later tales, those of Mitford's knight errant, adapted the might of Hercules for a later stage of society, one with different anxieties: What are the limits of power and the responsibilities of the powerful? How does property pass from one generation to the next? What do I owe my neighbor? And what does he owe me?[25]

What a brilliant thesis. Understanding myths as living forms continually reworked enabled Thirlwall to find in these tales the evolving sensibilities of antiquity. The myths surrounding Hercules recorded not what Hercules had done, but what the Greeks believed he must have done: how else would the world have come to be the way it is? "Hercules the greater" carries us back to the imagination of the ancient Greek who sought to comprehend the founding of community, to the days when fields had been cleared, boulders cleaved, rivers channeled to their present beds and wild animals slain. "Hercules the lesser" carries us back only as far as the threshold of civilization, to the days when society sought justice, propriety, the righting of wrongs and the relief of oppression.

In short, so mighty were Hercules's accomplishments that the ancients saw that he was no mere mortal and made him a god. So wide-ranging were Hercules's accomplishments that Thirlwall saw that he was no mere mortal and made him a conceptual device.

Nowhere does Thirlwall suggest this was a matter of allegory. Allegory is a self-conscious art form, composed at once on two levels. Not so ancient myth. But neither did these tales recount historical events. Their true meaning lay not in deeds, but in the forms of historical understanding they reveal and the changing sensibilities and concerns of both the storytellers and their audiences.

Thirlwall's study of mythical heroes was not limited to Hercules. Theseus, who was to Athens what Hercules was to the rest of Greece, similarly met the evolving needs of the ancient imagination. We find him slaying monsters and ancient beasts. To these exploits we may add his famous voyage to Crete and the slaying of the Minotaur, which brings us forward somewhat to a more settled day when Athens freed itself from sub-jugation. Finally, continued Thirlwall, we come to the tales of a less fabulous aspect but not greater historical likelihood. These speak of political reforms affected by Theseus as ruler of Athens.[26]

Minos, ancient king of Crete, affords a third example: a victorious prince and com-mander of a great naval empire, associated with the founding of Crete by Phoenicians, and also a wise and just lawgiver. Indeed, his reign was later seen as a golden era when legal and political institutions were first established that would later be associated with the Dorian and Spartan peoples.[27]

The task before the historian was to become a kindred spirit, forging a bond of sym-pathy and calling forth the genius buried within text and time. Sound history was not a matter of discrete facts, for such "facts" could not be separated from their context. The meaning of a tale was neither the genealogy it recorded, nor the migration it traced. It did not lie in the clearing of a road, the changing of a river's channel, or the righting of political wrongs. It lay in the issues addressed. Myth was the effort to comprehend how the world had come to be the way it was. It was not history (events); it was history (discourse).

A buoy is a marker. That footnote of Thirlwall's marked a profound change in how to read ancient myth.

THE AUTHORITY OF THE ANCIENTS

Thirlwall denied the direct authority of the ancients by reading them within the con-text of their understanding of the processes of history. A powerful example of this concerned pre-Hellenic Greece. Mitford's discussion rested upon the received opinion of the ancients that Greece had been overrun by barbarian hordes, the most prominent of which had been the Pelasgians. These peoples, spiced by colonies from Egypt and Phoenicia, had formed the background for the coming of the Hellenes and the Heroic Age of legendary Greece.[28]

Thirlwall's analysis is very different. He examines various ancient sources and mod-ern conjectures. Moving back and forth between past and present, he leads the reader to a vantage point that takes in both a broad view of the ancient landscape and the shapes of historical conjecture. The result is a certain distance between the reader and ancient Greece. What we seem to experience most directly is Thirlwall's judgment, his weighing of opinion, his insight. With Mitford we are drawn into the world of heroic deeds and events of great proportion. With Thirlwall that world is farther away, and we are drawn, instead, to wonder at the range and creativity of historical criticism. This is that essential fourteen-sidedness of things. The more we consider, the more complex and intriguing it all becomes.

The shift from fact to perception changes how one reads a text. Consider the genealogies so carefully reported by the ancients. Historians had long devoted elaborate care
over these often conflicting lineages. The "begats" of ancient literature, sacred and profane, framed a flow of events which involved migrations, the development of languages,
the passage of inventions and traditional rituals. Genealogy framed the prehistory of
mankind from the flood to the dawn of historical times, the age of Hesiod and Homer,
and of the writing of the Old Testament. In that article of 1832, Thirlwall observes that
the "usage of tracing the names of cities and nations to individuals . . . may have arisen
simply from the natural proneness of the Greeks to seek everywhere for persons who
afforded an object for the imagination to deal with, in the room of abstractions."[29]
This lovely suggestion that the Greek imagination was more comfortable with persons than abstract notions of development would be amplified in his *History* when he
considers the genealogies associated with King Pelasgus. Thirlwall observed that the
many grandsons of this most ancient King are each reported to have been the founder
of a city. Thirlwall saw in this that the inhabitants of these ancient cities and regions
perceived in one another a "natural affinity" that could only be explained by common descent. They saw themselves as cousins. Ancestral lines, like the mighty arm of
Hercules, were a conceptual device. Lists of begats were not historical records; they
were explanations of how different peoples could be alike in language, social custom,
or religious practice.[30]

Genealogies did not point backward in time, but horizontally across the surface of
ancient political and cultural relations. The Greek landscape is as rugged and mountainous as any in Europe, lending itself to a fragmented set of settlements cut off by
land from one another. And though there were connections by sea, villages and town-
ships developed distinct local customs and flavor.[31] This heightened the matter of affinities and the many variant lineages offered in myth presumably reflected the changing
landscape of perceived sympathies. They were not historical records of a deeper past,
but glimpses into the reasoning of the record-keepers themselves.

Thirlwall closes this section of his study by looking at the coming of the Hellenes.
He dismisses the traditional notion of a conquest by Hellen and his sons "on such a
subject, the authority of the best Greek writer is of very little weight."[32] But how can
anyone writing thousands of years after the fact have a sounder read of such a central fact than the ancients themselves? Thirlwall explained that the ancients lacked an
adequate understanding of historical agency. They abbreviated the complexities of the
gradual emergence of new social structures and values into a military conquest. The
long-supposed fact that the rise of Hellenic society had been the result of a military
conquest is stripped of its objectivity. It was how the ancients had seen the world, not
how the world had been.

This is the core element in Thirlwall's new conception of myth: it is how the ancients
had seen the world, not how it had been. Moreover, Thirlwall carefully demonstrates
that this understanding was itself embedded in time. Myth reworked its themes. In this
way it records the steady adaptation of the past to meet the demands of the present, an
adaptation framed by the reconfiguration of conceptual devices to meet new matters
of concern.

FIGURE 3.1 John O'Connor's *Pentonville Road*, 1884. Courtesy of Museum of the City of London.

Such reconfigurations are not unique to the distant past. As I write now I pause to look at a poster on my wall (see figure 3.1). It is of a late nineteenth-century London scene, painted by John O'Connor, showing the shops and row houses of Pentonville Road, with horse-drawn omnibuses and carriages. Looming in the background of figure 3.1 is a majestic gothic cathedral. Only upon closer inspection it turns out to be St. Pancras, one of London's great railroad stations: a new cathedral for the Industrial Age.

Implicit in Thirlwall's thesis is the notion that history is always about present sensibilities and present needs: Conceptual devices meet new demands, whether it's how fields came to be cleared, the appropriate form of a railway station, or how we can make sense of the changes in the world about us as traditional patterns give way to the engines of change of the industrial age.

FROM POETRY TO NATURAL RELIGION

Thirlwall's new mythos shifts profoundly the way ancient texts should be read. There is an echo here of a theological principle of interpretation known as the principle of accommodation, a lovely example of which is found in the reconciliation Galileo offered for the seeming contradiction between the Copernican hypothesis and the Old Testament. It is true that God tells Joshua that he will make the sun stand still, but, argued Galileo, this does not mean that the sun revolves around the earth, but rather that God appreciated that Joshua would have thought that was the case. Had God said he would make the earth stop spinning, Joshua would not have understood him. It was clearly not a good time to explain the mechanisms of the solar system; Joshua, after all, was in the middle of a battle, and so God accommodated his message to the sensibilities of his audience.[33]

Thirlwall's approach to ancient myth applies this same principle. We need to read the text in light of the unique, sustaining genius of its epoch and the pressing existential concern it addressed. The existing stock of constructs, the might and wisdom of a hero, kinship, migrations—these were not facts, but explanatory devices. And like having the sun stand still, they should be seen as what satisfied the imagination within a given epoch. The deepest aim of history, therefore, is not a chronology of events, but the developing sensibilities of a people witnessed in the recasting of ideas according to the genius of subsequent epochs.

It is fitting that we close with Thirlwall's discussion of the natural, as opposed to the later formal religion of the ancient Greek. Here are the deepest roots of Thirlwall's study: the foundations of mythos in the natural spiritual experience of the earliest Greeks.

To begin, Thirlwall objected to a vein of thought traceable to Herodotus, who had spoken of foreign input, chiefly Egyptian, in the development of Greek mythology. Since then historians had elaborated on his account, especially across the Enlightenment which favored the hypothesis that myth derived from the teachings of a priestly elite which had been cast within a "veil of expressive symbols and ingenious fables."[34] In time, however, these hidden teachings had been lost, leaving only their fabulous shell. It was this shell, blended with various local flavors, that formed the confused array Homer and Hesiod had organized and made more coherent through their creative genius.

This view, like the matter of the coming of the Hellenes, invites sharper lines than those Thirlwall believed characterize history. It opted for external plantings rather than internal growth, discrete invention and migration rather than internal, independent development. It was an understanding of history which failed to penetrate the surface: "They thought they held the roots of history, but they held in fact only dry twigs."[35] The true root of the tree was the unassisted imagination of the ancient Greek. Whatever similarities one might discern between Egyptian gods and those of Mt. Olympus, or between the legends of Phoenicia and those of Greece, these would lead the historian astray, if he did not begin with the genius of the ancient Greek itself.

Thirlwall begins with the root-level generalization that nothing to the ancient Greek was inert: "In all the objects around him he found life." The ancient Greek stood in awe of the bold forms, abrupt contrasts, and wonders of a "mountainous sea-broken land."[36] Nature and the wonder at her might were the very heart of the mythological mind.

About a decade later, Thomas Carlyle would give a series of lectures at the Royal Institution, lectures on heroes and hero-worship. The first of these was about the pagan, and Carlyle offered that pagan religions worshipped nature with "a recognition of the forces of Nature as god-like, stupendous personal Agencies, as Gods and Demons. Not inconceivable to us. It is the infant Thought of man opening itself, with awe and wonder, on this ever-stupendous Universe."[37]

Carlyle's lecture rings out harmoniously within the theater of Thirlwall's mythos. For Carlyle, as for Thirlwall, mythology was so central to the ancients that to reduce it to priestly quackery, poetic exaggeration, or allegory was to take the life out of it. Myth was serious belief. It was the effort to comprehend the mysteries of existence and to find a way to plant your feet and brace yourself as you looked upon the universe.

FROM NATURAL RELIGION TO CHRISTIANITY

Myth was serious belief, addressing the needs of its audience. What about Thirlwall's *History*? Did it, too, address the needs of its nineteenth-century audience? To answer this key question we need to go back to Thirlwall's essay introducing his translation of Schleiermacher's study of the gospel according to Luke. Thirlwall had come back from his year abroad so charged by the scholarly insight and wisdom of Niebuhr, Schleiermacher, and others that he set about bringing them before the English-speaking world. Yet he was sharply criticized for going beyond orthodoxy. What was this about?

The heart of his introductory essay concerned the nature of divine inspiration, the action of the Holy Spirit, and how we should read the gospels. This is a most nuanced topic and Thirlwall's piece is more than a hundred pages long, but perhaps we can adequately capture its drift with a few of his observations.

First, Thirlwall notes the now long-abandoned notion that the gospel writers were mere instruments which the Holy Spirit used to compose the gospels. Then, with characteristic irony, he added that when he says abandoned he means by the learned. The great majority, no doubt, still took it that the gospels were miraculously written by the Holy Spirit acting though Matthew, Mark, Luke, and John.

This hypothesis, however, was compromised by notable discrepancies between the three synoptic gospels of Matthew, Mark, and Luke, leading to a revised notion of a more limited inspiration. The gospel writers composed the gospels themselves, sometimes using their own judgment, but on key matters there was an intervening or superintending guidance. Most passages were still seen as divinely inspired, but others had been composed without the guidance of the Holy Ghost. This shifted the problem to how one might distinguish the two sorts of passages and so determine which bits are the inspired authoritative bits and which the merely human efforts. Though hypotheses were offered, they were all found wanting and a new approach had arisen, especially in Germany. This view saw the gospel writers as everywhere divinely inspired, but the inspiration was genuinely mediated through the individual writer. This approach shifted things once again, and this is where Thirlwall found Germanic scholarship so inspiring.

The gospel writers were historians, drawing upon their own experiences, stories they had heard, and written documents, including, perchance, the gospels of others. This set of material led to a host of different hypotheses on the part of modern commentators about the differences in the several accounts. But there was another matter here. Inspiration was not what happened when the Holy Spirit overrode the judgment of the gospel writers. It was *their* judgment. It all came down to their sensibilities, the way they saw things, their insight and wisdom. The Divine Word had become the wise word with a Christian sort of wisdom.

While one could have abandoned altogether the notion of a Holy Spirit, this new approach suggested we rethink what divine inspiration is really about: ridding ourselves, wrote Thirlwall, "of the exaggerations with which it has been loaded, and which were not implied in the judgment of the primitive church when it fixed the canon." He then proposed that we see the Holy Spirit "not in any temporary, physical or even intellectual changes wrought in its subjects, but in the continual presence and action of what is most vital and essential in Christianity itself."[38] That is, the gospel writers were everywhere inspired, but we should see this inspiration as stemming from the spirit within the writers and their sense for what was truly sacred in the teachings of Jesus.

More orthodox believers saw this as undermining the authority of the revealed word, but Thirlwall was really affirming the essence of Christianity and its sustaining truths. Here was the way to anchor the continued relevance of the Revealed Word to the modern world. Thirlwall is suggesting that if we could connect the Timeless Word of God to our times by reconfiguring its message in ways that resonate with our own pressing needs, then we would know how to draw comfort and guidance as we faced an uncertain future.

The authority of the story rested *not* on what was written, but on the spirit of things behind it. This is precisely the shift we have seen in Thirlwall's alternative to Mitford. It was not a matter of the facts you could glean from the legends of antiquity—even with the sharp reasoning that could transform the magical Golden Fleece into rivers rich in gold. It was rather to read the tale for the problem, the deep question it addressed, and so tease out the real "story" that was being offered.

We may add that it was common to charge this new German biblical criticism as applying the historical methods of the profane to sacred literature. But what is clear in Thirlwall is that it was just the opposite. He saw the profane as sacred—the likes of Hesiod and Homer as Matthew, Mark, and Luke-like—in that myths were serious literature and we needed to push to the underlying sensibilities of the storyteller.

IN THE END

Thirlwall's scholarship is brilliant. Perhaps we can take a page from his approach and ask ourselves if we can discern the real story that he was offering. What issue was behind his work, what deep question was he addressing?

We may go back to the years of Thirlwall's fellowship. The first thing he does is head off to the continent. This is not unusual in and of itself, but where the stereotype was to seek wisdom of a worldly sort, Thirlwall finds himself swept up by the brilliance of German scholarship. Returning home, he seeks a career which won't get in the way of contemplating "the finest products of the human intellect." There's a little of this and a little of that and then he settles on sharing his enthusiasm for German historical criticism; he translates one of Schleiermacher's essays. But the gap between his English readers and Schleiermacher is not just language. He must give the readers a fuller context, so that they may see what it is really about.

How frustrating it must have been to have been greeted with a chorus of criticism. He had found a treasure and they saw it as fools' gold . . . or worse. He took it hard. He might have gone about his business, settled into a legal career and carried on with his studies. But that would now no longer suffice. The message was too close to the messenger.[39]

He left law and returned to Trinity College. A project begins to emerge, an underlying push to all his subsequent work. He teams up with Hare to translate another brilliant piece of German historical scholarship, a secular work on the history of ancient

Rome, rather than an explicitly religious piece. But even here he was castigated for the implicit heterodoxy of Niebuhr's work. Niebuhr's *History of Rome* was harshly criticized in the *Quarterly Review*. As Niebuhr himself put it, this review "pronounces it a crime that clergymen of the Church of England should have translated a book containing the most disgusting scoffs at religion that have been written since Voltaire's time." Niebuhr and Thirlwall had evidently disturbed deep waters.

At this point Thirlwall withdraws strategically. He and Hare establish the *Philological Museum*, where they can write essays and reviews that expose the English reader to German scholarship—the "mythos" footnote was from a review of a work by Friedrich Kruse. And soon he shakes off the explicit link to things Germanic, and composes his own history of antiquity, a history of Greece. Thirlwall's approach differs dramatically from the prevailing English classical studies. He makes Mitford's *History* seem a relic of a distant past. Thirlwall's contextual read shifts from the ostensible tale itself to the pressing issues it addressed. Only by reading the tale through this lens can we understand what it was really about. The sustaining truths of mythology were not facts or events, but the judgment of the storyteller; just as the sustaining truths of the New Testament were not particular matters of fact but the judgments of the gospel-writers. Understanding the past is about the mind of the ancient storyteller, whether you are reading the Old or New Testament, or the tales of Hercules and Theseus. It is the wisdom of the storyteller that is sustaining.

In the end, it was always about the Holy Ghost and the continued relevance of an intellectual–spiritual read of life, even in an age of mass-produced straight pins, steam-driven looms, and scientific societies.

NOTES

1. Thirlwall, *Letters Literary and Theological*, l, 32, 45. See also J. Thirlwall, *Connop Thirlwall*, Clark entry in *Dictionary of National Biography*, and a lovely piece in Clark, *Old Friends*.
2. Thirlwall, *Letters Literary and Theological*, 53.
3. Thirlwall, *A Letter to Thomas Turton*, 20. See also Anon., "Charges Delivered by Connop Thirlwall," 291 and 298.
4. Mill, "Notes on the Newspapers," 592; see also Mill, *Collected Works*, vol. vi, 260.
5. *Penny Cyclopaedia*, vol. viii, 317.
6. Thomas, "Quite Early One Morning," in *Collected Stories*, 291.
7. Morgan, "Bishop Thirlwall," 80–81. We might note in passing that Thirlwall's quick mastery of Welsh might not have been as endearing as one might imagine. The Anglican Church sat in Wales only slightly more comfortably than it did in Ireland, and most congregants probably preferred English.
8. Stanley, *Sermons*, 217.
9. Plato, *Phaedrus*, 274B–278B.
10. Anon., "Charges delivered by Connop Thirlwall," 315–16, referring to an article in *Contemporary Review* of October 1875.
11. I first met Grote in Popper's *The Open Society and its Enemies*, a work first published in 1950; see also Arnaldo Momigliano's "George Grote and the Study of History," and Demetriou, *Brill's Companion to George Grote and the Classical Tradition*.

12. Grote resigned from the board of the University of London in 1830; see Robertson, "George Grote," 287. The Lafayette episode is discussed in Grote, *Personal Life of George Grote*, 60–62 and more fully in Grote, *Memoir of the Life of Ary Scheffer*, 37–48.
13. M. Clarke, *Greek Studies*, 38–39, 101.
14. Sedgwick, *Discourse*, 31–32.
15. Sedgwick, *Life and Letters*, vol. i, 336.
16. Froude, vol. i, 310.
17. Arnold, *Lectures on Modern History*, 110.
18. Mitford, *History of Greece*, vol. i, 19.
19. *Ibid.*, 90–100.
20. *Ibid.*, 30–31.
21. Thirlwall, "Kruse's *Hellas*," *Philological* 1 (1832): 322–23.
22. Plato, *Phaedrus*, 275D–276A.
23. This etymological digression draws upon the *Oxford Universal Dictionary*, a product of the Philological Society, whose founding membership included both Thirlwall and Grote. Further, Thirlwall was the society's first and longest serving president, from 1842 to 1868. See Marshall, "History of the Philological Society." See also J. C. and A. W. Hare, *Guesses at Truth by Two Brothers*, in such passages as this one: myth "owes its origin, not to the thoughts and the will of individuals, but to an instinct actuating a whole people: it expresses what is common to them all: it has sprung out of their universal wants, and lives in their hearts" (223).
24. Mitford, *History of Greece*, vol. i, 25–26.
25. Thirlwall, *History*, vol. i, 142.
26. *Ibid.*, 147–53. Thirlwall's interest in the unassisted imagination of antiquity is analogous to Niebuhr's development of the ancestral ballad thesis, where similarities in form meant the item was an expression of the mythological genius, but when the item was unique it was taken to reflect a particular event. See, among a host of fine works: Momigliano, "Perizonius, Niebuhr, and the Character of Early Roman Tradition," reprinted in *Essays in Ancient and Modern Historiography*, and Levine, *Dr. Woodward's Shield*.
27. *Ibid.*, 159.
28. Mitford, *History of Greece*, vol. i, 19.
29. Thirlwall, "Kruse's *Hellas*," *Philological* 1 (1832): 327, 350.
30. Thirlwall, *History*, vol. i, 44.
31. Grote develops this point in his careful examination of the geography of ancient Greece in the opening chapters of Part II, "Historical Greece," as opposed to "Legendary Greece," *History of Greece*, vol. ii.
32. Thirlwall. *History*, vol. i, 92.
33. Galileo, "Letter to the Grand Duchess Christina," *Discoveries and Opinions of Galileo*, 214. This is a fascinating aspect of biblical criticism; see both Forbes, *The Liberal Anglican Idea of History* and Kummel, *The New Testament*.
34. Thirlwall, *History*, vol. i, 210–11.
35. Hare and Hare, *Guesses at Truth*, 55–56.
36. Thirlwall, *History*, vol. i, 208–09.
37. Carlyle, *On Heroes*, 17.
38. Thirlwall, "Translator's Introduction," xix in Schleiermacher.
39. See Clark, *Old Friends*, for a charming account of the young Thirlwall's stay in Italy (95–98) and his interview with the prime minister when he was offered the bishopric of St. David's which opened with a discussion of Schleiermacher (132–34).

4

A BANKER'S SON AND
A PHILOSOPHIC RADICAL

As we turn from Thirlwall to Grote, we turn from Cambridge to London, from the university to the city. These two places offer rich and richly contrasting images—the cloister and the market place. There is Trinity College with its magnificent Wren Library, fine ancient buildings, and ancient traditions walled in medieval isolation. All owned by the Church, not simply by deed, but by practice. And there is London, with its omnibuses, whose children are Oliver and the Artful Dodger. The London buildings are different, the air is different, and it was all owned by commerce.

But Grote's London is less his material setting than his intellectual setting. Take the year 1818. Thirlwall graduated from Cambridge that year, and Grote, whose formal schooling had ended eight years before at the age of sixteen, was working at his father's bank in the city. It was then that the economist, David Ricardo, introduced the young Grote to James Mill and so to the Benthamian "school." In this school we have the proper complement to Thirlwall's Trinity College. This was a matter of kindred spirits and the intimacy of table talk—James Mill would for years be a weekly dinner guest of the Grote's. It was reading circles—Grote belonged to a small group, including John Stuart Mill, which met early in the morning to discuss leading works by Ricardo, Mill, and others. It was, as well, the special role that Bentham's many manuscripts played in developing a set of common analytical approaches. And it was a matter of joint projects, such as the founding of the University of London and of the *Westminster Review*, agitation for political and legal reform, and indeed, Grote's career as an MP.

We can all conjure up the image of the scholar working at his table, alone with his imagination, as in Alan Lightman's delightful *Einstein's Dreams*—yet we should not underestimate the value of such networks as Grote's in London and Thirlwall's at Trinity College. It has been shown, for example, that your chance of winning a Nobel Prize increases dramatically if you have worked in a lab with someone who has already won one. Such groups raise the intensity of the give and take of notions, and each profits from the intelligence of the whole.[1]

Grote was born in 1794. He went to the Charterhouse School at the age of ten, where Thirlwall and Hare were his classmates. He was there for just six years, at which point his father, having little regard for formal schooling, sent him to work at the family bank. George was the eldest of the Grote children, destined to assume the mantle of the banking business, especially as his father had become very keen on hunting and the life of a gentleman. From then on Grote was self-taught, taking advantage of his spare time to read and study. We learn from his journal, for example, that on Sunday, March 28, 1819 he "rose at ½ past 5. Studied Kant until ½ past 8, when I set off to breakfast with Mr. Ricardo. Met Mr. Mill there, and enjoyed interesting and instructive discourse with them, indoors and out (walking in Kensington Gardens), until ½ past 3, when I mounted my horse and set off to Beckingham."[2]

In 1815, Grote met and fell in love with Harriet Lewin. The duplicity of an acquaintance who also sought the hand of Ms. Lewin set the courtship back. This was somewhat of a "young Werther" moment, and Grote took it hard. He did not press his love, but withdrew, wrapped in the folds of his loss. In time the lie was uncovered, George approached Harriet, and the relationship flourished again, only to be stopped virtually in its tracks when George's father firmly expressed his disapproval.[3] George was able to convince his father that with a trial period he might change his mind. After their engagement had been put on hold for almost two years, however, the father was still opposed. The young couple took themselves off to a small church to get married, telling no one for a month.

As a dutiful eldest son, George had agreed to leave school and work at the bank. He had agreed to wait for his father's approval before marrying the woman he loved. But there was a limit to his father's prerogative. We can only imagine what sort of reception the young couple received when they announced they had wed, but the new couple set up home in Chelsea and George continued to work at the bank. George's mother seems hardly to have figured in all this, having already thrown herself into an evangelical movement. There could hardly have been a greater gap between parents and child than between this young Benthamian and his Tory father and his devoutly religious mother; in a letter of 1823 he referred to his home as overclouded by the "deepest night of ignorance."[4]

George was fully captivated by the vivacious Harriet, by her love of music, and the whirl of their social life. But hardship continued to track them. Harriet gave birth prematurely in January of 1821 and the baby died within a week. On top of this loss, Harriet then nearly succumbed to an attack of puerperal fever. Sadly, this was an all too familiar tragedy. It would be another quarter of a century before Ignaz Semmelweis, an Hungarian physician, would demonstrate that the incidence of puerperal fever and related childbirth deaths would be dramatically reduced if doctors followed disinfection practices like routinely washing their hands in chlorinated lime.

The Grotes would not have any children.

In the world of ideas Grote was a utilitarian, and in his working life, a banker. Surely here was a practical man of affairs, placing reason above faith, utility above spirit, armed with a critical calculus fit for any occasion. Yet in his private life he was the very model of a romantic. Dutiful and earnest, Grote's path was strewn with conflict and he responded bravely and with passion. When he could not go to university, he woke early to carry on his studies. He put up with his father's objections to marriage for close to two years, but then went ahead without parental blessing. After the loss of their baby, the young couple, guided by Harriet, threw themselves into the arts and various projects. Sadly, even here, there were traps. Harriet discovered a singer whom they took under their wing. But the young woman proved to be wild, became pregnant, and left her daughter to be cared for by the Grotes. They formed a deep attachment to the child, who spent much of her first four years with them. Occasionally the mother would return for the child only to abandon her again, and then finally she took her away. The Grotes were devastated.[5]

Grote was a private soul and his wife his only confidante—perhaps a consequence of having grown up in a household where his views on life, the universe, and everything else would have been highly suspect. Like Thirlwall, we know the Grotes would go for long walks. In a letter from Harriet to Murchison we learn that Murchison and Lyell were favorite guests of the Grotes, and that they would go out for informal geologizing together, but it was less that reflective ramble and commune with nature that was so important to Thirlwall and more a matter of good talk. With this, perhaps, is a basic quality that distinguishes our two classical scholars. Learning for Thirlwall was at the very heart of life; recall that he looked forward to a quiet life "contemplating the sights and sounds of nature and the finest productions of the human intellect." But there is not the same contentment for Grote. His learning was most often yoked to a cause or an argument—to a case he was making. He may have read Col. Sleeman's *Recollections and Rambles through India*, but it was less out of curiosity than to further support his comparative study of mythology.[6]

His scholarship was exhaustive, his *History* richly annotated, and much the longest of the three accounts of legendary Greece in our study. It excels in its brilliance, in the drama of the march of mind, as he chronicles the growth of critical thinking from early tentative qualms about the literal truth of the ancient tales to the reasoned judgments of Aristotle or Strabo. He assembles his evidence to demonstrate a clean six-sided figure. It is not that things are not complicated, but rather that he pushes through to a clarity the complexities had masked.

As Thirlwall flourished in the quiet quads of Trinity College and the seclusion of St. David's and Abergwili, so Grote flourished in the busy day of banker and man of affairs, with associates who were economists and lawyers, as well as scholars.

What a marvelously matched pair these two are; how richly do they span the currents of this early Victorian era. Grote, the young Benthamian, met to discuss the tracts of the new social philosophers, as Thirlwall dined with Sedgwick, Whewell, and Hare, and hosted visits from Coleridge. Banker and Broad Churchman, the one a founder of the Reform Club, the other a guiding light at the founding of the Apostles.[7] They seem almost to cover Victorian life and letters between them, suggesting the deep influence John Stuart Mill wrote of in his essays, where all thinking Englishmen are either a little Coleridge-ian or a little Bentham-ian, whether they thought about politics, poetry, or philosophy.

At the same time, we ought not to carry this continental divide too far. These many scholars knew one another, were familiar with one another's works, and most often held each other in high regard both intellectually and personally. Grote, Sedgwick, and Lyell were each members of the Athenaeum. There are many such facts which weave together the lives of these scholars. Sedgwick, Whewell, Hare, and Grote were among the founding members of the Philological Society, whose first president was Thirlwall. Both Thirlwall and Hare were fellows of the Geological Society, and

Whewell and Grote corresponded over aspects of Plato's physics, a matter which may have arisen in talk at "the Club," as they were members of Sam Johnson's literary club, whose mid-Victorian membership also included Murchison and the historians Macaulay and Milman.[8]

WITH MOUNTING INTENSITY

In the spring of 1830, the Grotes cleared a space in their busy lives and headed off to the continent. As we saw earlier in the drawing room of the Athenaeum, this trip was marked by the excitement of France's July Revolution, where Louis Phillippe replaced Charles X on the throne.

When they returned to London, they were quickly immersed in a swirl of events— so much so that Harriet would write in her journal in December her husband was being pulled in too many directions. With the passing of his father, George, as executor, had been burdened with its many complexities. Further, the many anxieties over the state of the country and agitation for parliamentary reform had heightened the demands on him at the bank. Nevertheless, she noted, he had managed to add several chapters to his "History" over the preceding five months. Then, late in January, James Mill pressed George to write again on parliamentary reform. He had written a pamphlet back in 1821, and Mill thought the time was ripe for another push. This pamphlet would prove to be most successful and led to people urging him to put his name forward as a candidate for Parliament. Harriet worried that this was premature. The literary renown certain to derive from his "History" would be crucial in sustaining his political career: "The 'History of Greece' *must* be given to the public before he can embark in any active scheme of a political kind."[9]

The pressure on Grote to run increased across 1831. He met with the elder Mill and maintained his decision not to join the fray. As late as November of 1831, Harriet noted: "The History draws ahead, and I trust will continue in progress steadily through the ensuing winter."[10] But Grote gets caught up in the tumble of events surrounding the Reform Bill. In June of 1832 Bentham dies, the Reform Bill passes, and Grote announces his decision to stand for election.

The voting would take place in December, but he had to interrupt both his politicking and his history-writing for that remarkable occasion we discussed earlier: his appearance before the secret Parliamentary committee on the condition of the Bank of England.

No wonder Harriet was so concerned about the demands on her husband's time. Nor, clearly, did things lighten up after the intensity of that June. When the polls closed in December, Grote led the polling with 8,788 votes, thereby becoming one of four Members of Parliament for London in the first reformed Parliament.

These were heady days. Optimism ran high among the reformers. A host of issues and reforms people had long sought suddenly became possible—expanding the electorate, factory laws to protect the worker, even establishing national public schools. In the excitement, Harriet notes that her husband set the history project aside, so swamped had he become with his many duties.[11]

AS AN MP

As an MP Grote jumped right into things, taking on a striking array of matters. His maiden speech concerned the complexities of municipal corporations and taxation. A few days later in March of 1833, he spoke against the Irish Coercion Bill. Though the problem of unrest in Ireland was real, he argued that replacing courts with military tribunals and denying both the right of assembly and the right to petition was far too harsh. In the coming days he also seconded a bill for a national education scheme and brought the House's attention to the ongoing struggles in Portugal and their impact on trade. But in all this, the issue that commanded his attention above all others was the secret ballot.

In the swirl of causes and debates the real promise of the reformed Parliament for Grote lay in a core reform he had called for in both his pamphlets. Parliamentary reform was more than getting rid of the absurdity of pocket boroughs. That was just a beginning. He had written "elections by the people were a pure *fiction*." And that was still the case. Members of Parliament represented those who had elected them, and this was but a fraction of the people, less than 4 percent.

Even here, the influence of the elite was magnified by the mechanics of the electoral process.

Polling in England had been public for a long time; one might say it was of a time where the memory of man runneth not to the contrary. One stood forward and announced his vote. For most of this time virtually everyone worked for one member of the landed elite or another; thus to exercise independent judgment and vote for a candidate in opposition to the lord's choice took considerable courage. In his pamphlet, Grote refers to the outrageous instance of the Duke of Newcastle actually expelling tenants who voted for a candidate who was not his choice. To this Grote added: "Without secret voting there cannot be public-minded voting; and without public-minded voting, men worthy to be legislators can neither be singled out nor preferred scarcely even created."[12]

On April 25, 1833, Grote delivered a moving address calling for the secret ballot. He began with the words of Lord John Russell, when he had introduced the Reform Bill with a view of Parliament as it ought to stand: "So constituting this house, as that it should enjoy, and command, and deserve, the confidence of the people." For this to be the case, Grote assured his colleagues, required an expanded electorate and the privacy to exercise their judgment unhindered by the pressures of the ruling classes. He addressed a number of issues and closed with this powerful sentiment: "Above all, you are called upon to make this House, what it professes and purports to be, a real emanation from the pure and freespoken choice of the electors; an assembly of men commanding the genuine esteem and confidence of the people, and consisting of persons the fittest which the nation affords for executing the true end and aim of government."[13]

The bill failed 211 noes to 106 ayes.

Grote renewed his efforts five more times in the coming years, but the secret ballot would not become standard practice in England until 1872.

The failure to win the secret ballot was not the only setback for the reformers. In the first draft of his *Autobiography* John Stuart Mill rendered a strikingly harsh judgment: "Nobody disappointed my father and me more than Grote, because no one else had so much in his power. We had long known him fainthearted, ever despairing of success, thinking all obstacles gigantic; but the Reform Bill excitement seemed for a time to make a new man of him: he had grown hopeful, and seemed as if he could almost become energetic. When brought face to face however with an audience opposed to his opinions, when called on to beat up against the stream, he was found wanting."[14]

We have glimpsed a certain reticence in Grote, back when he was first courting Harriet and again when his father disapproved of their getting married. And we may understand why he might have been hesitant to act directly, given the lack of sympathy between his world view and that of his parents. But it is a long way to the timidity suggested by the term "fainthearted." For sure, Grote was no Lyndon Johnson. He did not work the crowd, nor seek to manipulate his fellow politicians. Yet he was active and earnest in pushing a reform agenda, and he did so with a strong-hearted regard for the power of reason and the virtues of his causes.

In fact, Mill did not include this charge in the published version of his memoir. Perhaps he changed his view of Grote's faintheartedness, or perhaps he decided it would be too hurtful. It is more likely, however, that he changed his mind after he himself had served in Parliament and saw how daunting it is to face an audience opposed to one's opinions—or worse, to face their indifference. We learn from a note that Thirlwall went to the House of Commons to hear a speech by Mill and how saddened he was to see every seat of the front bench of the opposition (the Tories) was empty.[15]

In the published *Autobiography* Mill does not mention Grote, but we can perhaps faintly hear his name, as he wrote instead: "And now, on a calm retrospect, I can perceive that the men were less in fault than we supposed, and that we had expected too much from them . . . It would have required a great political leader, which no one is to be blamed for not being, to have effected really great things by parliamentary discussion when the nation was in this mood."[16]

Grote remained in Parliament until the elections of 1841, but he grew increasingly disenchanted, remarking that "it is not at all worth while to undergo the fatigue of a nightly attendance in Parliament for the simple purpose of sustaining *Whig* Conservatism over *Tory* Conservatism."[17] It was time to clear his study of the distractions of both Threadneedle Street and Westminster, of both the bank and politics, and to throw himself into his scholarship.

The remains of Thirlwall and Grote lie buried in the same grave at Westminster Abbey.

THE MARCH OF MIND

Grote had first conceived of writing a history of Greece in 1822, long before Thirlwall's *History* would appear, and we may gather from his wife's comment that he had made considerable progress before he set it aside late in 1832. As with Thirlwall, Mitford's very successful *History* was the provocation for Grote's work.

Mitford, a country gentleman, had been a Tory Member of Parliament, and his *History* vigorously critiqued the democratic city states of antiquity, linking them directly to contemporary events here at home—as in this passage about the mob rule of the Athenians, which would not be "strange among ourselves, where county meetings, too frequently, and the common hall of London, constantly exhibit perfect examples of that tyranny of a multitude."[18] These criticisms were not lost on his audience. In an early essay in the *Westminster Review*, Grote replied to Mitford's enthusiasm for anything other than democracy by enumerating atrocities executed by ancient Greek oligarchies and then added testimony of indisputable weight, lest anyone think this had not been by design. He quoted an oath cited by Aristotle as formally sworn by some of the oligarchies, containing these words: "I will be evil-minded toward the people, and will bring upon them by my counsel whatever mischief I can." What a remarkable governing principle, and what makes it even more remarkable is Aristotle's own modest reprimand: "Let them misgovern if they choose; but let them at least employ some decent pretences to delude the people into a belief to the contrary."[19]

If it strikes you as curious that anyone would look at the Golden Age of Athens and find its democratic impulses as fundamentally flawed, that is because Grote's defense of Athenian democracy carried the day—and the week, and the centuries. Grote's *History* dramatically transformed the story of ancient Greece, forging the modern framework where Athenian democracy is the centerpiece of the Greek legacy.[20]

He came to this view naturally. By 1832 he was a leading "philosophical radical." This did not mean that "the people" would never make mistakes. As he explained in a letter of 1823, "the people are right upon the long run—right more frequently than they are wrong—and above all, that they have no interest in going wrong." This notion of interest is what we saw earlier in *The Handbook of Political Fallacies*. We can think of it as a structural tendency that comes out of given political systems. So, Grote continues, monarchies and oligarchies "have a permanent and incurable interest in plundering and depressing the people in order to gratify their own appetites for wealth and power, and therefore however wise *they* may be, their wisdom will never be applied to the benefit, but to the injury of the people."[21]

A NOTE ON HISTORICAL CRITICISM

Such a complete flip in historical judgment reminds us of a rather curious conundrum. We tend quite naturally to think of the past as fixed; after all it has already happened, and the future as open. But it's rather more the other way around.

At the heart of history is the notion of forces pushing and pulling events along; that there is some sort of agency or structural tendency that gives shape to the passage of today into tomorrow. The computer, social media, and technology more broadly is one such force often on the modern mind, as are structures like class and religion. The future is not so open.

What about the past? How could it be open, if it's already occurred? History is what we make of history. This is the bold print of Thirlwall's brilliant read of myth. It's not what myth said that was important, but what it addressed. What was on their minds. What they were trying to explain.

Seeing history as an account of "how the world came to be the way it is" means the past is being read both in terms of the historical forces that give us the "coming to be"—the mighty arm of Hercules—and the pressing concern on the historian's mind. These matters are open. And because they are open, they lend themselves to a vision of what we might do to make things better. For Mitford, Athenian democracy was mob rule of the sort you see in public gatherings. That Athens lost its way in its epic struggle with Sparta becomes a key lesson for his day. For Grote, however, Athens had nurtured civic virtue and brought out the best in its citizens—a goal England could accomplish, if it would but try. The past is less a fact than an argument, an argument that serves as a guide to the perplexed.

It would be a mistake to think that politics is only found in such matters as constitutions or public meetings. Historical scholarship is not neutral and then put to political ends. It emerges from sensibilities deep in the world view of the scholar. These several histories by Mitford, Thirlwall, and Grote were serious politics, but what made them so was their foundation in historical criticism. Grote's approach to the earliest chapters of the Greek experience, to the meaning of myths, will as rightly lead us to the foundations of his politics and his formative debt to the Benthamian "school" as his defense of Athenian democracy or his views on parliamentary reform and the secret ballot.[22]

The key is the concept of fiction.

Myth was fiction. They each agreed on this. For Mitford this had meant the legends of antiquity had been distorted by poetic exaggeration. The task of the historian was to scrub away at these tales with a rationalist's scouring powder, reducing fancy to fact. For Thirlwall the key feature of myth as fiction had been the resonating sensibilities of poet and audience. Drawing upon a stock of conceptual devices, the poet addressed the deepest questions and concerns of the day. Myths were not idle entertainment. They carried the sensibilities of a people: the meaning of life, the universe, and everything else. They were the divine inspiration of their age. The task before the historian was to uncover this deeper meaning through an inspired leap of imagination.

What did fiction mean to Grote? He begins with the simple observation that fiction is not bound by the actual, nor beholden to the plausible. Its authority rests in its ability to satisfy the soul. Before Grote, from antiquity on through to the nineteenth century, it had been assumed that these legends had some historical basis. For Thirlwall, the greater plausibility of the political Hercules, Theseus, or Minos did not imply a greater historical probability. Yet even he had said of the Argonautic expedition: "The tradition must also have had an historical foundation in some real voyages and adventures."[23] Grote did not deny that there was more or less matter of fact in the mix; only that we could know what it was. In a discussion of the Eleusinian mysteries, for example, he urges that it would be "impossible to ascertain and useless to inquire" what was fact and what fancy, for belief in these tales did not stem from their approximation to real fact.[24]

Criticism could not separate the wheat from the chaff. The door is closed to the history (events) within myth. Grote did not credit any myth with more or less historical validity, going so far as to apply this even to the Trojan War. There was nothing behind the curtain. Whatever history might be reflected within myth was beyond the critical gaze of the historian.

As Grote explained, the boundary between fantasy and fact did not announce itself. Galen, after all, had not known whether Asklepius had been a god. This is even more complicated for the modern reader. As offspring of the critical sensibilities that would develop in ancient Greece, we instinctively read these tales as we would poetry, extracting metaphorical truths. We are too much accustomed to matters of fact to conceive of a time when these beautiful fancies were "construed literally and accepted as serious reality."[25] Even so, Grote adds, we are not immune from myth making impulses ourselves! Here Grote treats us with an intriguing tale of modern mythology fit for a Harlequin novel: a tale of murder and torment in the streets and elegant palazzos of modern Florence.

Grote's tale pulls together several works—an essay by Goethe, a biography of Lord Byron, and a novel by Lady Caroline Lamb—but its real beginning is Byron's dramatic poem "Manfred." Here are a few lines from this haunting poem. It will warm up our imaginations as we watch the tormented hero, Manfred, conjure forth spirits to help him:

SPIRITS: What wouldst thou with us, son of mortals,—say?
MANFRED: Forgetfulness—
SPIRITS: Of what—of whom—and why?
MANFRED: Of that which is within me; read it there—Ye know it, and I cannot
    utter it.[26]

Manfred's life has fallen apart and he beseeches the gods to allow him to forget his lost love. Bare fragment that this is, it suggests the issue Goethe would take up: what torment had Byron shared with his fictional Manfred? Goethe finds Byron's poem so powerful, it must have derived from some wrenching experience. It could not simply have been imagined. What had caused Byron to suffer so, that he could write so movingly?

Goethe finds an answer in a horrid occurrence which had haunted Byron for years:

When a bold and enterprising young man, he won the affections of a Florentine lady. Her husband discovered the amour, and murdered his wife; but the murderer was the same night found dead in the street, and there was no one on whom suspicion could be attached. Lord Byron removed from Florence, and these spirits haunted him all his life after.

Goethe has drawn this extraordinary tale from a novel by Lady Caroline Lamb, whose central character resembles Byron. Unfortunately, this marvelous tale of Byronic man has nothing to do with Byron. Thomas Moore, who had written a biography of Byron, assures us that Goethe's tale has no basis in fact. It was, to quote Moore, "an amusing instance of the disposition so prevalent throughout Europe to picture Byron as a man of marvels and mysteries, as well in his life as in his poetry."[27]

Here, writes Grote, is a genuine legend "such as even now, in the age of Blue Books and Statistical Societies, holds divided empire with reality." Goethe's romance "is not a mis-reported fact: it is a pure and absolute fiction. It is not a story of which one part is true and another part false, nor in which you can hope, by removing ever so much of superficial exaggeration, to reach at last a subsoil of reality."[28] Like Goethe's tale, ancient myth was not drawn from historical experience, but emerged full-blown from the poet's imagination. It was a past which had never been "a region essentially mythical, neither approachable by the critic nor measurable by the chronologer."[29] Here is the clarity of a cube.

This being said, one would imagine that Grote's study of "Legendary Greece" would have been little more than a wink and a nod, but it spilled across thirty-one chapters and filled more than 680 pages. Grote's skepticism is complete, but it was no retreat from the evidence. It was a problem shift.

From the get-go, Grote broke the mold. His approach to the history of legendary times was not chronological, but topical. He started with myths chiefly about the gods, and then surveyed the legends regionally: such as those from Argos, Arcadia, and Crete. Still other chapters were given to special topics, like the Trojan War. And most strikingly, the final chapters offer a comparative study of mythology and religion, an aspect of his approach evident from its earliest days, as can be seen in a letter from January, 1823: "I am at present deeply engaged in the fabulous ages of Greece, which I find will require to be illustrated by bringing together a large mass of analogical matter from other early histories, in order to show the entire uncertainty and worthlessness of [such] tales."[30]

There is a further nuance here. Not only is myth fiction, but it was also serious belief: "The popular faith, so far as it counts for anything, testifies in favour of the entire and literal mythes, which are now universally rejected as incredible."[31] That is, not only did myths need never have mirrored experience, but they had been accepted literally. This is the special place for Grote's comparative study. Here, for example, is a passage he cites from Col. Sleeman's discussion of Hindu beliefs: "The Hindoo religion reposes upon an entire prostration of the mind,—that continual and habitual surrender of the reasoning faculties, which we are accustomed to make occasionally, while engaged at the theatre, or in the perusal of works of fiction."[32]

But why had Grote examined ancient mythology so carefully, if it is so tenuously tied to experience? Because it had been the proverbial clay in the ancient potter's hands. The reworking of these legends in the commentaries of later generations gave shape to a new form of thinking—critical thinking, and that is the tale Grote tells. Every peoples have had a body of myth, tales of heroes and gods, of ancestors and their struggles against forces of gigantic proportion, but only in ancient Greece did these tales provoke a searching criticism. Only in ancient Greece did myth give way to analysis, fiction to nonfiction. Mitford had used ancient legend to track the progress of civilization after Noah's flood. Thirlwall evoked ancient sensibilities, capturing the richness of the ancient imagination. Grote traced the emergence of the modern mind. As he announced in his preface, his task was "to set forth the history of a people by whom the first spark was set to the dormant intellectual capacities of our nature."[33]

THE THREE FACES OF CRITICISM

Legendary Greece stretched across a tumble of centuries, from the earliest mythical storytelling in the dark recesses of the distant past, across the heyday of Hesiod and of Homer, and on into historical times. Along the way many poets and playwrights reworked these tales and commentators weighed their meaning. Some worried about discrepancies in the tales or wondered just how such events could have happened. In their efforts to reconcile and reconstruct these legends something happened: open credulity gradually gave way to criticism, and this criticism grew into the Golden Age, the age of Thales and Anaximander, of Socrates, Plato, and Aristotle, of Herodotus and Thucydides.

Let's sample Grote's account, starting with that central body of legend, the tales surrounding Hercules. Hercules was a great problem for the earliest critics. Grote uncovered a delightful example in a note on the Kalydonian boar hunt. This had been one of the great aggregate dramas of ancient Greece, along with Jason and the Argonauts, the siege of Thebes, and the Trojan War. Ephorus, who wrote in the third century B.C., wondered why Hercules had not participated. He decided that he must have been engaged in another adventure at the time.[34] This rather sweet apology is fully predicated on the reality of the mythical world in all of its detail. And detail is the key here.

This understanding was typical of one group of commentators. They were compliers, reconciling different accounts and straightening out inconsistencies. A closely related body of criticism was guided more directly by changing moral sensibilities. This is a fascinating body of criticism, and Grote provides a host of instances. Pindar, in the fifth century B.C., repudiated several myths, such as when he denied that Tantalus would have slain his son, Pelops, and then served him to the gods. This so outraged Pindar that he argued it must be due to a slanderous enemy (something a Hatfield would have said of a McCoy?).[35] Regarding Hercules, there is an instance from the later critic Apolodorus, in the second century B.C., who felt it deeply unjust that Hercules is said to have avenged himself against the sons of Neleus, while the father, the actual offender, escaped with his life. So he claimed this passage in the *Iliad* was spurious.[36]

In an earlier day such alterations would have occurred organically in the retelling of myths, according to Thirlwall, as the poet reflected and gave shape to the changing sensibilities of the community. By the fifth century B.C., evidently, we have moved from the age of the rhetor to that of the author, from a wandering bard singing the tales of the ancients and changing them as he went along, to the critic offering commentary.

There is yet another mode of criticism, with a lovely instance concerning the legends surrounding Hercules. Amidst his wanderings and adventures, Hercules comes to the home of Admetus, an old friend. He finds him disconsolate over the recent death of his wife, Alkestis. It turns out that Admetus had been gravely ill and Alkestis had bargained with death, the god Thanatos, swapping her life for the imminent death of her husband. Hercules then calls out Thanatos himself and challenges him to a wrestling match, and in defeating death, Alkestis is restored. Grote observes that this marvelous tale was subsequently rationalized to far more prosaic proportions: Plutarch, in the first century A.D., simply refers to Hercules's medical skills, as illustrated by his cure of the deathly ill Alkestis.[37]

This last brand of commentary Grote calls historical. In surveying the commentaries of Herodotus, Thucydides, Pausanius, Strabo, Plutarch, and others, it becomes clear what set them apart was their jaundiced eye. Their skepticism led them to question particular items, despite a general regard for the mythical past. This might lead to a certain discomfort as is evident in a passage from Herodotus in the fifth century B.C. concerning the incredible exploits of Hercules. This time it is the occasion when Hercules is said to have slain a thousand men with the jawbone of an ass. While it is easy enough to see how an ass's jawbone would be a lethal weapon, Herodotus finds it hard to imagine how one could slay a thousand men. Even putting Hercules at some strategic narrow pass, after a while there would be such a vast mound of the slain that the next victim would have to climb for a long time to get his chance to fight Hercules. Herodotus then adds: "I pray that indulgence may be shown to me both by gods and heroes for saying so much as this."[38] Herodotus clearly felt he had ventured onto thin ice. Plutarch, writing in the first century A.D., is more confident as he prays that the fabulous material surrounding the life of another hero, Theseus, "may be so far obedient to my endeavours as to receive, when purified by reason, the aspect of history."[39] Reasoned purification is precisely the tale Grote sought to trace.

There was a vitality to all of this criticism, and on this Grote echoes both Carlyle and Thirlwall. To us these tales are intriguing fancies; children's stories, but to the ancients they were serious belief. Once it was realized that they could not be taken as they presented themselves, the effort to figure out what they were really about became the leading edge in the growth of critical thought.

The tale Grote tells is a noble one. Mankind had begun to emerge from the open credulity of the mythological mind. The march of the intellect had taken its first steps. Two elements were crucial here. One was the move from an oral culture to the written word. Writing allowed the legends to stand "apart from the imagination and the emotions wherein the old legends had their exclusive root."[40] There was, as well, "the habit of attending to, recording and combining, positive and present facts, both domestic and foreign."[41] Such facts isolated the outrageous proportions of myth, and guided its rationalization.

The play of such present facts was subtle. For many, the mythical past had been real, but it had not been an ordinary reality. To borrow a lovely metaphor from Pindar: the present was only half-brother to the past. But this led to a tension between faith and reason.[42] For example, Diodorus, in the first century B.C., protested against those who reduce the deeds of ancient heroes to the proportions of life as we know it. Applying the common measure of humanity to such figures lacked grace. The ancients required their own standards.[43]

The progress from credulity to criticism was fused by Grote with the progress from religious faith and superstition to science. There was a fundamental opposition between science and religion, what Grote termed "the radical discord between the mental impulses of science and religion."[44] The facets of this progress are nicely illustrated in commentary on the legend of Deukalion, a legend with many parallels to the story of Noah's flood. Unremitting rains sent by Zeus to punish the corrupt laid the whole of Greece under water except for its highest mountain tops, where a few souls, led by Deukalion, found refuge. The devastated world was repopulated, the legend continues, by Deukalion and his wife, Pyrrha, who cast stones over their heads, the stones turning into men. Justin, in the second century A.D., denied the miracle of casting stones into people by casting Deukalion as a king who gave shelter to fugitives of a great flood.[45] Aristotle also commented on the legend of Deukalion. He denied the religious character of the tale by denying the flood had been a divine judgment against a wicked race; the flood waters had been the result of periodic atmospheric cycles. There was little room for Zeus's anger here.[46]

Aristotle's comment is a perfect segue into the next phase of our study, on the sciences, but we are not quite ready for it. Grote saw both Justin and Aristotle as testing the past against a standard set by the present, a testing which left no room for faith. Such "present facts" lie at the heart of the emergent rational criticism of the historian. One of the most distinctive aspects of Grote's approach was his use of extensive comparative materials on the mythologies of various peoples. Grote's notebooks, in fact, include an array of materials on the mythology of the American Indian, of India, and of Europe. They illustrate the extent to which Greek mythology was but an instance of the universal condition of the precritical mind.[47]

THE EMERGENCE OF MODERN SOCIETY

The scholarship we have just traced was classical studies in its heyday. Latin, Greek, and the history of antiquity were central pillars of schooling. The radical discord between Mitford's approach to the earliest chapters of ancient Greece and those of Thirlwall and Grote is stunning, greater even than their political differences. It is as if we had stepped from a horse-drawn carriage onto a locomotive. Thirlwall and Grote transform Mitford's scholarship into a relic of a distant past almost quaint in its naiveté.

Yet the data, as it were, is the same for each of them. They had each read the same set of ancient writings. Such is the charm of a scholarship resting on shards of a distant past, a situation that is just the opposite of Strachey's image of a vast ocean of material for the nineteenth century. Clearly the historical judgment of Thirlwall and Grote was commanding and their histories brilliant. But, we may ask: Was this brilliance itself a product of historical forces?

Somewhere in the writings of Bertrand Russell is a proof that the intricacies of the individual exceed an infinite intersection of universals. There is always a further touch of particularity. So we cannot derive these scholars or their scholarship from historical forces or conditions. But perhaps there is a prompt, a provocation, common to the age, that helps us see why this work at this time. Both Thirlwall and Grote link storytelling to pressing issues of the day. For Thirlwall such issues are what myth addresses; for Grote it is what commentary explains. We have seen that there was a host of such issues in 1832, stemming largely from industrialization. Transformation was visible. The traditions and practices of the past no longer offered sufficient guidance. The question was, where was it all heading?

For Grote, what made the ancient writers so compelling is that critical reason itself was in the process of being forged: the iron was still hot and the likes of Herodotus and Thucydides, Parmenides and Plato wrote with the excitement of discovery. A similar quality explains the distinctive interest we continue to take in the early nineteenth century. It is fresh with an excitement surrounding the press of change . . . and with it distinctly modern sensibilities.

The early nineteenth century is a watershed. There is a break in the landscape and things tumble into modern times. In 1832 the push for reform brought people into the streets. Tens of thousands marched peacefully to demand political change in what has become the hallmark of political movements through the civil rights and anti-war demonstrations of the 1960s on to Tiananmen Square in 1989, the Arab Spring early in the twenty-first century, and over a million people in Paris protesting terrorism more recently. As we looked more closely at the push for reform we could see a broader shift from the countryside to the city, from the landed estate to the factory. Industrialization changed not only how people worked, but where they lived and the connection between the worker and his master. The lord of the manor had, out of time immemorial, been the collective father—charged with husbanding the collective resources. When it came time to vote, of course one would vote publically. It was an opportunity to swear allegiance to the father. When the workplace shifted from field to factory, however, clerks who spent their days surrounded by other clerks had no such allegiance. There was no collective father. We may recall John Stuart Mill's observation that men would be held together by new ties, "for the ancient bonds will now no longer unite, nor the ancient boundaries confine."[48] Hence the special significance of Dickens's Scrooge, who had to be taught to step forward as the new father. The new working men had to make their own way, but there was also place for those with resources to take care of those who worked beneath them. Tellingly, though Whigs supported parliamentary reform, both Whigs and Tories represented the old patriarchy and resisted the secret ballot until it was clear the father had died.

It is not just politics that takes us back to this era as a watershed. There is literature. Though there had been poetry and novels before, there is a marked change with the Romantic poets on the one hand, and the work of Walter Scott and Jane Austen on the other. They are somehow modern, and we read them without feeling someone needs to explain them to us. To some extent this may be a matter of language and remote cultural references, but more profoundly there is a shift, a shift that is up front in both Thirlwall and Grote. It is a shift from event to perception. Story becomes a window to an interior landscape. For Thirlwall this is a matter of deciphering what was really on the mind of the ancient bard as he told his tale and engaged his audience. It was about the storyteller's imagination. For Grote it was about the critic and the issues that led him to modify the ancient tale and make it more authentic or reliable as history. This "new" text allows us to trace the real glory that was Greece—more than the poetry of Homer, the sculptures of Phidias, or the armies of Alexander, it was the growth of criticism. Every peoples have had a body of legend handed down from generation to generation, but only in ancient Greece was that body of storytelling tugged at, stretched, trimmed, and laid out flat so that reason was satisfied. Canons of criticism emerged which led to axiom and precept, to postulate and principle, and Plato and Aristotle took the place of Achilles and Odysseus as guides to the perplexed. For the modern reader, the beginning of modern storytelling is less what people do than what they are thinking as they do it; just ask Henry James or Freud.

In *The Discovery of the Mind*, Bruno Snell argues that Homeric figures had no internal dialogue, no "self" as it were, and further, that we can follow in subsequent Greek storytelling the gradual development of modern sensibilities about the mind.[49] That is, Homer tells us what people did, not what they were thinking as they did it. There was no inside. With Thirlwall and Grote that is both affirmed and turned on its head. There is no outside to ancient legend, no report on historical events. What we really see is perception, not deeds, be it Thirlwall on the imagination or Grote on criticism.

We may imagine that centuries ago some soul saved the last copy of one of Plato's dialogues from a fire, or perhaps he left a set of notes on Aristotle's lectures on religion or humor or some other topic—we don't know because they were left to decay, abandoned in a collapsing house. In either case, no one has had a greater impact on classical studies since then than George Grote. He upended the table, shifting the gaze of history to the glory that was Athenian democracy. His study of mythology, the first book of his *History*, set the stage for this shift by showing us how we should take the measure of what the Greeks had taught us in critical thought and discourse.

Our focus on Grote's analysis of myth and the march of mind has had a different end: to set the stage for the nature of historical criticism at this moment—a complement to Thirlwall's approach. Together they represent the very model of the new criticism of their day.

The nineteenth century witnessed the rise of the sciences. By the close of the century science had become the authoritative voice in western culture, the laws of thermodynamics replaced the ancient constitution of Athens, and Aristotle became more widely known for the views rejected by the scientific revolution than for the merits of his philosophy. And so we, too, pass from classical studies to science. We shall be examining the work of two master scientists: the Rev. Adam Sedgwick, Woodwardian Professor of Geology and fellow of Trinity College, Cambridge; and Charles Lyell, who in 1831 had written the first volume of a most influential text, *The Principles of Geology*. Intriguingly, we will find that the cutting edge of early nineteenth-century geology mirrors the developments we have just traced in classical studies.

NOTES

1. Robertson, "George Grote," in *Dictionary of National Biography*; see also Clarke, *George Grote*, and H. Grote, *Personal Life of George Grote*. Regarding networks, see Bruner, *Culture of Education*, 132, 154. See also Menand's *The Metaphysical Club*, which examines the work of several leading American scholars in the latter half of the nineteenth and early twentieth centuries, and see especially the analysis of Charles Sanders Peirce's difficulties when he lost his place in any network (435–36).

2. H. Grote, *Personal Life of George Grote*, 37.

3. Clarke, *George Grote*, 10–25, and Richardson, "A Regular Politician in Breeches," 136–37.

4. Grote, Grote papers at University College London, letter of February 25, 1823; cited in Fuller, "Bentham, Mill, Grote," in *Brill's Companion*, 126.

5. Eastlake, *Mrs. Grote*, 87–88.

6. George Grote Papers, British Museum, Ms. 46126, entry 344. The letter is dated February 8, probably 1864. Mrs. Grote invited the Murchisons, adding that the Lyells and the Milmans were also invited, and that he should bring "Hammers tongs and a' [presumably a way of rendering 'all' in an accented manner]" for a bit of local geologizing. Further, Grote's notebooks include an array of materials on the mythology of the American Indian, of India, and of Europe. They illustrate the extent to which Greek mythology was but an instance of the universal condition of the precritical mind. See holdings of British Museum and the University of London collection.

7. Grote was among a small circle of Whigs and Radicals in 1836 who formed a club to help foster and maintain the energies of the Reform movement. Even the celebrated chef at the Reform club, Alexis Soyer, was a social activist. Among other projects he wrote *Soyer's Shilling Cookery for the People*, aimed at the poor and middle classes; see Kurlansky, *Choice Cuts*, 118–22. In *Cambridge Before Darwin*, Garland discusses the influence Hare and Thirlwall had with a small group of undergraduates who formed a "social-cum-discussion-group called the Apostles" (64). This group included F. D. Maurice, Tennyson, Sterling, and Monckton Milnes.

8. Cowell, *The Athenaeum*, and Ward, *History of The Athenaeum*. Thirlwall enjoyed the famed breakfasts of his friend Richard Monkton Milnes, where one could chat with Lyell, Grote, Disraeli, or Baron Bunsen, in Thirlwall, *Connop Thirlwall*, 158–59. Regarding Johnson's Literary Club see Timbs, *Clubs and Club Life in London*, 184.

9. H. Grote, *Personal Life of George Grote*, 67.

10. *Ibid.*, 70.

11. *Ibid.*, 75.

12. Grote, "Essentials of Parliamentary Reform," in *Minor Works*, 12.

13. Grote, "Parliamentary Speech of April 25, 1833," in *Minor Works*, [25] of Bain's introduction.

14. Mill, *Autobiography, Early Draft, Collected Works*, vol. i, 203.

15. Thirlwall, *Letters to a Friend*, 83.

16. Mill, *Autobiography, Collected Works*, vol. i, 204–05.

17. Grote, *Personal Life of George Grote*, 127.

18. Mitford, *History of Greece*, vol. ix, 60–61.

19. Grote, "Fasti Hellenici," *Westminster Review* 5 (1826): 290; see also Grote, *Minor Works*, [15] of Bain's introduction. The passage is from Aristotle's *Politics*, bk 5, ix, Grote's translation; see also *Aristotle's Politics and Poetics*, 143.

20. Kierstead, "Grote's Athens"; Momigliano, *Studies in Historiography*.

21. Grote, "Letter of 1823," cited by Kierstead, "Grote's Athens," 174–75.

22. Momigliano saw Grote's *History* and his politics as linked this way: "Nothing was more important to the Philosophical Radical than the careful evaluation of evidence." This was so, Momigliano went on, not only in their legal studies and their political and social theories, but also in their historical writings. Momigliano, "George Grote and the Study of Greek History," reprinted in his *Studies in Historiography*, 62. We will return to this link in our study of Bentham in chapter 7.

23. Thirlwall, *History*, vol. i, 165, 147, 149, 170–71.

24. Grote, *History*, vol. i, 42 and similar passages on 48, 51–52, 207, 239, 416–19. In "Grecian Legends and Early History," Grote wrote: "That there is more or less matter of fact among these ancient legends, we do not at all doubt. But if it be there, it is there by accident" (132).

25. Grote, *History*, vol. i, 332, 345–56.

26. Byron, "Manfred," 812.

27. This material comes not from Grote's *History*, but from "Grecian Legends," 80. Goethe's original review appeared in *Kunst und Alterthum* in 1820, and Byron passes it along to his friend Thomas Moore as something he would find interesting; see Byron and Moore, *Works of Lord Byron*, vol. iv, 320–25 and vol. xi, 71–73. It is worth noting that Goethe's tale bears a close resemblance to Lady Caroline Lamb's widely read and scandalous novel of 1816, *Glenarvon*, written after her affair with Byron had come to an end.

28. Grote, "Grecian Legends," 81–82.

29. Grote, *History*, vol. i, 42.

30. H. Grote, *Personal Life of George Grote*, 41.

31. Grote, *History*, vol. i, 414.

32. *Ibid.*, 414 (see note at bottom of the page).

33. *Ibid.*, v (from Preface).

34. *Ibid.*, 142.

35. *Ibid.*, 154.

36. *Ibid.*, 110 (see note at bottom of the page).

37. *Ibid.*, 113 (see note at bottom of the page).

38. *Ibid.*, 381.

39. *Ibid.*, 200.

40. *Ibid.*, 353; see his lengthy quotation from M. Ampere's *Histoire Litteraire* in a note on pp. 346–47, and also Eric Havelock's brilliant discussion of oral culture in antiquity in *Preface to Plato*.

41. *Ibid.*, 351.

42. *Ibid.*, 364.

43. *Ibid.*, 398.

44. *Ibid.*, 362.

45. For the reader intrigued by the connection between stones and living things, see Mircea Eliade's fascinating study of fire and early religious sensibilities, *Forge and the Crucible*.

46. Grote, *History*, vol. i, 97.

47. Among the works Grote examined: the writings of Lewis and Clark, Bancroft's *History of the United States*, and Catlia's *Letters on the North American Indian*; Col. Sleeman's *Rambles and Recollections*, James Mill's *History of India*, H. H. Wilson's *System of Hindoo Mythology*, and Colebrooke's *Essay on Sanskrit*; Turner's *History of the Anglo-Saxon*, Fauriel's *Sur les Origines de l'Epoque Chevalrique*, and Middleton's *A Free Inquiry*. Grote's many notebooks may be found in the British Museum, Senate House Library, and the University Library at Cambridge University.

48. Mill, *Collected Works*, vol. xxii, 228–29.

49. Snell, *Discovery of the Mind*.

5

THE VICTORIAN IDEA OF SCIENCE

Toward the end of a workshop many years ago, the historian Arnold Thackray was asked how he would characterize early Victorian science. Thackray replied by painting a picture of Adam Sedgwick giving an open lecture on the shore of the North Sea outside Newcastle on Tyne. Thackray was right on target here, but there are other candidates he might have turned to as well.

JOHN HERSCHEL AND LARDNER'S CABINET CYCLOPEDIA

The close of the eighteenth and the beginning of the nineteenth centuries witnessed the rapid growth of the London scientific community, a growth which brought into prominence a newly professional scientist. The eighteenth century had been the age of the patron, whether we speak of letters or politics, pamphlets, plays, or parliamentary seats. The Church of England was part and parcel of this system, as were Oxford and Cambridge. But it was not only a matter of jobs; there was also the relationship between patron and the artist. Though the Church was the leading supporter of science in the eighteenth century, it was as patron and not as employer. A cleric's pursuit of natural philosophy was an acceptable, even laudable enterprise, but it was not a responsibility.

As we turn into the nineteenth century we find scientists by vocation. This scientist was hired by any of a number of new institutions. The Royal Society of London had been the only scientific society until the close of the eighteenth century, yet by the 1830s there were some ten additional societies, such as the Geological Society founded in 1807. There were also new schools such as the University of London as well as the London Mechanics' Institute, which would later become Birkbeck College, and also the Royal Institution, a research facility.

The importance of these developments is the presence in London of a large group of scientists. But the changes were not just a matter of numbers. Something else changed, to the extent that several scholars have suggested this period be seen as a second scientific revolution, one where the style of argumentation became more sophisticated mathematically, richer in data and statistical analysis. Further, the shift from gentleman-amateur to professional meant societies moved from a broad, cultural purview to a more narrowly technical and pragmatic practice.[1] Much was afoot.

One prominent figure in this swirl was John F. W. Herschel. While in an aristocratic society there would have been any number of children born with proverbial silver spoons in their mouths, there was only one of whom it might be said that he had been born with a silver telescope to his eye. John's father, William Herschel, and his aunt, Caroline, were both astronomers of the first order. Not only did they discover a planet, Uranus, the first planet to be discovered that you cannot see with the naked eye, but they went on to examine and catalog a host of galaxies, offering a bold theory as to what their shapes tell us. So impressive was this work that William was made astronomer to the King and the Herschel household moved from Bath to Windsor.

John, in fact, would prove to be an accomplished astronomer in his own right, spending four years in Cape Town, South Africa, cataloging the stars and nebulae of the Southern Hemisphere and so completing the project his father and aunt had started. But astronomy was only one of his interests. He was a leading proponent of the wave theory of light, a pioneer in photography, and active in the Geological Society.[2] Thus when Dionysius Lardner sought to kick off his publishing venture, the Cabinet Cyclopedia, which would include many volumes in the sciences, it was natural that he ask Herschel to write a volume on the nature of science, and so we have *A Preliminary Discourse on the Study of Natural Philosophy*.

One of the features of this era is the striking array of efforts by scholars to reach broader audiences. Take for instance the matter of encyclopedias. There were two notable additions to the already long-standing *Encyclopaedia Britannica*, first established in 1768: the *Encyclopaedia Metropolitana* (1817) inspired by Coleridge, who wrote an introductory essay on the method of the project; and the *Penny Cyclopedia* (1833), which could be purchased piecemeal in small tracts of thirty-two pages each. These two publications were actively supported by the scientific community. Herschel, for example, wrote lengthy entries on light, sound, and physical astronomy for the *Metropolitana*, and in a letter, William Whewell remarks on his Trinity College colleague, George Peacock: "I found Peacock busily employed in comparing the numeral words of the Nanticocks and the Mandigoes, and proving that people talked decimal notation in central Asia—all for the *Encyclopaedia Metropolitana*."[3]

It is clear from the rise of penny libraries and mechanics' institutes that the working classes and the rising middle class—recall Francis Place—were willing to set aside scarce resources for access to learning, be it books or lectures. Consider Dr. Dionysius Lardner's Cabinet Cyclopedia. Lardner had been the first professor of natural philosophy at University College. Stepping down from that post, he soon initiated his Cabinet project, which from 1830 to 1844 offered 133 volumes on various scholarly topics, chiefly in history and the sciences. For six shillings one could buy Thirlwall's *History of Greece*, Herschel's *Preliminary Discourse*, or volumes on mechanics, on hydrostatics, optics, chemistry, heat, and geology. It is worth adding that six shillings is less than half the usual price at this time for octavo-size books, with quarto volumes running at much steeper prices.[4] Herschel's *Preliminary Discourse* was the initial volume of the project, reflecting both the exalted place of the sciences in this era, and the stature of John Herschel.

The bulk of the *Preliminary Discourse* is a substantive account of the leading concepts and principles in the sciences. We need not follow this work in any detail, save to share a story Herschel tells in the introduction that captures a certain take on the essence of science.

There is a familiar line, but lovely still, from Pascal that the heart has reasons that reason knows not. We might form a parallel to this effect—that reason has its imaginings that imagination knows not. John Playfair, writing in 1807, put it this way: "We became sensible how much farther reason may sometimes go than imagination can venture to follow."[5] He was talking about the breathtaking change in the landscape when seen through the lens of James Hutton's new theory of the earth. An ordinary coastline, Siccar Point on the North Sea, was transformed into an "abyss of time" as Playfair came to see layers of rock represent a whole mountain range raised and then eroded back to the sea and its debris upended and raised again. Herschel is making the same point about the wonderful powers of scientific reasoning, but he does so with a more dramatic tale.

In 1820, the HMS *Conway* commanded by Captain Basil Hall left England on a voyage that took it to the Pacific, with various ports of call including the Galapagos Islands and San Blas on the west coast of Mexico. From there they sailed to Rio de Janeiro. The stretch from San Blas to Rio covers some eight thousand miles and took eighty-nine days, during which they did not sight land and passed only one other ship. Whatever the character of the first part of this voyage, rounding the tip of South America would not have been a pleasant sail, as it is marked by bitter cold, lashing winds, and treacherous seas. And so, to borrow from Captain Hall's account: "After crossing so many seas and being set backwards and forwards by innumerable currents and foul winds," the captain gave the order to "hove to" at four in the morning—that is, to slow the ship down to a slow drift.[6] You can imagine the crew's dismay at such an order. Cold and wet, out in the middle of the sea for months, they were no doubt anxious to get into port. Why would the captain pull up short in the middle of the night in the proverbial middle of the sea? But though they had not sighted land for months, the captain had always known where they were. He had reined the *Conway* in at four a.m. because he knew they were near Rio. As it happened, that morning proved to be most foggy and still the captain held back; when suddenly the skies cleared there they were with the great Sugar Loaf Rock, which stands at one side of the harbor's mouth, straight ahead.

This, for Herschel, was the very wonder of science: not a wonder at things, not a wonder at the colors and contours of Sugar Loaf Rock, but a wonder at how, across those three months and the relentless wash of the sea day after day, with bitter winds and violent currents and no land sightings to mark their place, the captain had always known where they were. It was a wonder at the reasoning that could yield so masterful a command of nature.

Thomas Carlyle, writing just a bit later that same year, gives us a different take on mankind's command of nature. In *Sartor Resartus* Carlyle evokes the traditional tale of the Wayland Smithy, pointing to a "little fire which glows star-like across the dark-growing (nachtende) moor." Asking if that fire is a thing unto itself, he replies that it was kindled at the Sun and fed by air that circulated before Noah's Deluge. Within that flame and forge: "Iron Force, and Coal Force, and the far stranger Force of Man, are cunning affinities and battles and victories of Force brought about; it is a little ganglion, or nervous centre, in the great vital system of Immensity." Then, changing his metaphor, Carlyle likens that fire to an altar, "whose dingy Priest, not by word, yet by brain and sinew, preaches forth . . . from the Gospel of Freedom, the Gospel of Man's Force, commanding, and one day to be all-commanding."[7]

Staying with Carlyle, in 1840 he began a set of lectures about heroes, the heroic, and hero-worshipping. In the first of these talks he comes back to this same wondrous near-visceral feel for nature, evoking the might of the pagan hero and the character of paganism. The first man who began to think, the first pagan thinker, "open as a child, yet with the depth and strength of a man"; what would he have thought? "Nature had as yet no name to him; he had not yet united under a name the infinite variety of sights, sounds, shapes and motions, which we now collectively name Universe, Nature, or the like,—and so with a name dismiss it from us. To the wild deep-hearted man all was yet new, not veiled under names or formulas; it stood naked, flashing in on him there, beautiful, awful, unspeakable. . . . This green flowery rock-built earth, the trees, the mountains, rivers, many-sounding seas;—that great deep sea of azure that swims overhead; the winds sweeping through it; the black cloud fashioning itself together, now pouring out fire, now hail and rain; what *is* it? Ay, what? At bottom we do not yet know; we can never know at all. It is not by our superior insight that we escape the difficulty; it is by our superior levity, our inattention, our *want* of insight. It is by *not* thinking that we cease to wonder at it. Hardened round us, encasing wholly every notion we form, is a wrappage of traditions, hearsays, mere *words*. We call that fire of the black thunder-cloud 'electricity,' and lecture learnedly about it, and grind the like of it out of glass and silk: but *what* is it?"[8]

There is a difference between words and things, between a wonder at understanding and the thing itself. In the *Preliminary Discourse* Herschel wrote: "We must never forget that it is principles, not phenomena,—the interpretation, not the mere knowledge of facts,—which are the objects of inquiry to the natural philosopher."[9] Herschel was on the side of words, and we agree with him here, but it is still the case that he is not the very model of early Victorian science. He lacked the passion we can feel so palpably in Carlyle. And what is more, though he was gifted and accomplished in many and diverse fields, others reached more effectively across society, such as our next candidate.

CHRISTMAS LECTURES AT THE ROYAL INSTITUTION

A second candidate for the very idea of science also comes from the physical sciences. Looking at England across the first third of the nineteenth century, two areas of study stand out: the new electrical sciences galvanized by the discovery of batteries and current electricity, and the science of heat with the first steps toward both a kinetic theory of heat and the laws of thermodynamics. In both of these we see a certain jostling, an image of the world beneath experience marked by busy-ness, replacing the more sedate and tempered image characteristic of eighteenth-century science, where particles were held by forces acting at a distance; each to its own place. As conceptions of heat came increasingly to be about heat as a mode of motion, matter came increasingly to be seen as vibrant within, even as its outward appearance would be calm. To this was added the emergent views of chemical activity with a myriad of electrical pushes and pulls at the atomic level. The rather contemplative state of eighteenth-century matter was giving way to something far more chaotic.

In a series of researches in 1831 and 1832, Faraday conducted almost four hundred electrical experiments. These were of such significance that he was awarded the Royal Society's Copley Medal in 1832.[10]

Some thirty years earlier Nicolson and Carlisle had been the first to put leads from a "voltaic pile," as batteries were first known, into water. Their discovery of bubbles of oxygen and hydrogen excited the scientific community. Electricity could tear water apart into its constituent elements. Over the intervening decades many performed such experiments, but there was little clarity about the array of results until Faraday's work. He effectively showed that the electricity from the leads was disrupting the electrical balance of molecules, creating relatively stable but unbalanced particles he called "ions." These ions would migrate across the solutions to the "opposite" terminal; positive ions to the negative lead, negative ions to the positive.

To explain the various effects of his experiments, Faraday abandoned the analogy with gravity—that an electrical force emanated from the leads the way gravity emanates from the sun. Instead, he pictured solutions as cluttered with lines of electrical force emanating not only from the leads but from each of the myriad ions. These lines of force interacted with one another, pushing and pulling particles hither and yon. James Clerk Maxwell would later give mathematical form to all this, creating field theory, but how apropos that Faraday had envisioned a molecular world much like the urban busy-ness of London's streets there outside the Royal Institution, just down the lane from Piccadilly Circus, in the metropolitan center of the world's first industrialized society.

If we fast forward to the end of the century, we find the French master physicist and philosopher, Pierre Duhem, contrasting two takes on theorizing in physics: abstract theories and mechanical models. And while continental scientists favor the former, the English have this peculiar fascination with the latter, a fascination Duhem traces back to Faraday. Duhem then adds, speaking of the English approach:

> In it there are nothing but strings which move around pulleys, which roll around drums, which go through pearl beads, which carry weights; and tubes which pump water while others swell and contract; toothed wheels which are geared to one another and engage hooks.
>
> We thought we were entering the tranquil and neatly ordered abode of reason, but we find ourselves in a factory.[11]

Michael Faraday had been born to a family of limited means and at a tender age had been sold off to apprentice with a bookbinder. Even though his employer was good and intelligent, we must picture a lad thirteen years old, working a fourteen-hour day and sleeping alone in a tiny room with few amenities. By dint of much study and hard work Faraday not only learned to read and write, but developed an interest in the sciences. He was given tickets to a series of public lectures by Sir Humphrey Davy, the great chemist and director of the Royal Institution. Sometime following these lectures, Faraday

summoned all his courage and approached Davy, presenting him with a lengthy manuscript. He had transcribed his notes on the lectures. Davy was suitably impressed
and hired the young man as an assistant. Faraday went on to an extraordinary career
in the new electrical sciences, performing experiments and making observations that
would lay the foundations not only for electrolysis and electromagnetic field theory,
but also for the curious phenomena whereby electricity and magnetism induce effects.
Intrigued at the prospect of harnessing these inductions, Faraday went on to develop
the electric motor. Here indeed was Duhem's physics on the factory floor.

John Tyndall, Faraday's successor at the Royal Institution, shares a charming episode long after Faraday had become a leading scientist. He and Faraday were leaving
the lab one evening, to visit a colleague. As they left, Faraday said, "Come, Tyndall,
I will now show you something that will interest you." Walking north, they

> reached Blandford Street, and after a little looking about he paused before a statio-
> ner's shop, and then went in. On entering the shop, his usual animation seemed
> doubled; he looked rapidly at everything it contained. To the left on entering was
> a door, through which he looked down into a little room, with a window in front
> facing Blandford Street. Drawing me towards him, he said eagerly, "Look there,
> Tyndall, that was my working-place. I bound books in that little nook."[12]

We can imagine Tyndall stunned into unaccustomed silence; such a small nook. He
might have wondered how Faraday looked back on his childhood. *Does he often think
of the nights he spent here? Is he saddened by the loneliness he would have felt then; or
does his rise from privation suffuse him with pride? No, I think not; more likely he says
to himself "There but for the grace of God . . ." This little room, a long day's errands,
fetching supplies, sweeping the pavement out front, and all the other incessant busy-ness.
And then, at the end of the day, or perhaps early in the morning before the start of the
next day, in any case by candle light and not sunlight he would sit and write out his
notes from Davy's lectures . . . some three hundred pages-worth . . . thinking through the
links that connect one observation to the next, capturing the idea and letting it run its
course; instead of capturing a flag and running across a field. Does he miss a childhood
he never had?*

How sharp the difference between the childhoods of Herschel and Faraday: Herschel
privately tutored and schooled at Eton and Cambridge, and the substantial inherited
sum of £25,000 from his father and further lands and property with the passing of his
mother; Faraday apprenticed at seven and self-taught. An essay written by Charles
Lamb in 1825 reminds us how great the cost of such a life.[13] It opens:

> If peradventure, Reader, it has been thy lot to waste the golden years of thy life—thy
> shining youth—in the irksome confinement of an office; to have thy prison days
> prolonged through middle age down to decrepitude and silver hairs, without hope
> of relief or respite; to have lived to forget that there are such things as holydays,
> or to remember them but as prerogatives of childhood . . . Melancholy was the
> transition at fourteen from the abundant playtime, and the frequently intervening

vacations of school days, to the eight, nine, and sometimes ten hours' a-day attendance at the counting house.

Founded early in the nineteenth century, the Royal Institution was an industrial age research facility charged with investigating various ends of value to its sponsors. It pursued the improvement of fertilizers, the miner's lamp and a host of other practical improvements. It also took seriously its charge to be a teaching institution, both within and without the sciences. One could attend Davy's lectures there and also hear Coleridge speak of Shakespeare or Carlyle speak of heroes and hero-worshipping. Later, when he became the head of the Institution, Faraday showed he had not forgotten his roots. He instituted a Christmas series of lectures for the poor children of London. One set of these lectures, a truly marvelous piece, *The Chemical History of a Candle*, is still in print more than a century and a half later.

Here was the new industrial age harnessed to the physical sciences. Power had shifted. The groom in the lord of the manor's stable would in a few decades become the mechanic of steam driven devices. The science of steam and combustion, of electricity and coils of wire wrapped about iron cores is far removed from the mathematical scholia of Newton's *Principia* or Herschel's efforts to measure parallax in the stars of the Southern Hemisphere. Yet this science of Faraday, while it reflected the dramatic growth of industry, is too much a matter of applied works to capture the idea of science in this era. Its reach is mediated by gear works and lines of force, and too rarely taps the soul or the greater reaches of the imagination. Let us turn to Adam Sedgwick and that beach outside Newcastle.

AN OPEN TALK FOR THE BRITISH ASSOCIATION

The best place to begin is the summer of 1838, when some three to four thousand coal miners stood on the beach at Tynemouth outside of Newcastle. They had gathered to hear a lecture by Adam Sedgwick. "I am told by ear- and eye-witnesses that it is impossible to conceive the sublimity of the scene," John Herschel wrote:

> as he stood on the point of a rock a little raised, to which he rushed as if by sudden impulse, and led them on from the scene around them, to the wonders of the coal-country below them, thence to the economy of a coal-field, then to the relations with the coal-owners and capitalists, then to the great principles of morality and happiness, and at last to their relation to God, and their own future prospects.[14]

Sedgwick, the much beloved professor of geology at Cambridge, was the formative intellect behind the concept of a geological system or period, still the way we structure the history of the earth: from the Pre-Cambrian to the Quarternary. What an extraordinary event this must have been: a master geologist giving an open air lecture to thousands of working men. How did this come about? Why was Sedgwick there? Why would he arrange to give a public lecture? And why would there have been such a gathering of workers?

On the Condition of the Working Classes

It is worth pausing to make what we can of the condition of the working classes at this time. As it happens we may draw upon two parliamentary studies: one led by Michael Sadler, MP for Leeds, on factories in 1832 and the other led by Lord Ashley, later the Earl of Shaftesbury, on the conditions of those who worked in the coal mines, most particularly women and children, in 1842.

The Sadler report records the testimony of men and women who were old and often deformed at twenty, after tending machinery twelve and even fifteen hours a day from an early age. One man, Joseph Hebergam, seventeen, described how things had been when he started working at the age of seven: "In the morning I could scarcely walk, and my brother and sister used, out of kindness, to take me under each arm, and run with me to the mill, and my legs dragged on the ground; in consequence of the pain I could not walk." The pathos of a seven-year-old becoming deformed by his work after six months and unable to walk the mile to the mill, is matched by the testimony of a father whose children worked in a mill. At the busiest times of the year their work day would begin at three in the morning and end at ten or half past ten at night. That's more than nineteen hours. The father added, "we have cried often when we have given them the little victualling we had to give them; we had to shake them, and they have fallen asleep with the victuals in their mouths many a time."[15]

Despite such testimony, no action was taken on Sadler's proposal to limit the working day to ten hours. Soon elections were called for the new reformed Parliament, and Sadler narrowly lost his seat to Macaulay. Lord Ashley was approached to pick up the cause. He agreed, and so began a life-long effort to reform the worst abuses of the workplace in Victorian England.

At the start of the chapter on the effort to reform the mines, the Hammonds excellent biography of Lord Ashley offers the stunning observation "that the complacency with which England accepted the social misery that accompanied the progress of the industrial revolution was very like despair." The roots of this despair were the findings of the dismal science. Economics had uncovered laws felt to be every bit as unflinching as those of Faraday's electrical science. England's economy was dependent on manufacturing, and manufacturing depended on capital. The lot of the worker was hard, but if you tried to ameliorate the conditions of the worker, you would drive capital away, throwing the factories into difficulties and workers out of their jobs. This economic determinism guided the views and deeds of those like James Graham, who was a minister in Lord Grey's government in 1832.[16]

It had taken the combined light of the Renaissance, the scientific revolution, and the Enlightenment to break the hold of traditional, medieval society. The ancient regime maintained itself as long as thoughtful people feared the complexities of change.[17] Such bold experiments as the United States, a government established on principles drawn not from practice out of time immemorial but from the writings and convictions of thoughtful men, rested on the conviction that if we would reason together, a better way would be found. Now the Industrial Revolution had made havoc with traditional society. Yet reason was shackled by its own success. We knew enough of life's complexities to doubt any significant improvement was possible, a sensibility not far removed from that of a much earlier day.

Into this context Ashley set the findings of his commission. To be successful, the staff of the commission would have to override intellectual misgivings with the emotional power of their portrait of the sufferings of the workers. They succeeded.

It was common, they found, for mines to employ children of seven. In many pits younger children, five and six years old were employed. The youngest were trappers. They were charged with repeatedly opening and closing doors to shafts that enabled fresh air to be drawn into the mines. A little boy or girl would sit alone in the darkness of a small cavity with a string in their hand for twelve hours or longer. Here is the testimony of Sarah Gooder, aged eight:

> I'm a trapper in the Gawber pit. It does not tire me, but I have to trap without a
> light and I'm scared. I go at four and sometimes half past three in the morning, and
> come out at five and half past. I never go to sleep. Sometimes I sing when I've light,
> but not in the dark; I dare not sing then. I don't like being in the pit . . . I would like
> to be at school far better than in the pit.[18]

Children also pushed small carriages of coal along passages leading out from the coal face; they might stand in water ankle-deep all day operating pumps or they might operate the cages that took the workers from the surface to the depths of the mine. In some few instances children were hired because they were small enough to handle the tight quarters, but for the most part it was a matter of pay scale. Men would require a salary of 30 shillings a week, while a child would earn but 5 or 7 shillings.[19]

The testimony pertaining to women was equally powerful. Men and women working together in the mines, almost naked, offended deeply felt notions of decency. Again, the primary motive here was the difference in salaries.

A bill to exclude all women and girls, and boys under the age of thirteen, was brought to the House of Commons, and passed virtually without opposition. But then it had to go to the House of Lords. Here there was much opposition. A leading critic of the bill, Lord Londonderry, suggested there were "superior advantages of a practical education in collieries to a reading education," and that boys were as fit to work in the mines at the age of eight, as at ten. Yet despite such sentiments the Lords evidently felt they could not throw the bill out, and satisfied themselves by amending it. The bill did pass, and was an important success, though there would be no mine inspectors until another act in 1850.[20]

Such was the deplorable condition of the working classes in the 1830s—but why would miners and industrial workers turn out by the thousands to hear a science lecture?

Science and Penny Tracts

The rise of industry caused major shifts in population. By the end of the eighteenth century London had over a million people and major industrial centers popped up in formerly quiet provincial towns like Birmingham and Leeds. The population of Bradford, for example, had grown by 50 percent every ten years from 1811 to 1851. In 1851 only 50 percent of the population had been born there.[21] This was more than a matter of growing disproportion in representation in Parliament. Traditional ways of life had been lost and new ones were being forged on the spot. Not always effectively; recall Dr. Baker's study of the cholera outbreak in Leeds in 1832.

The scientific community responded to the profound changes of industrialization, and not just with public health reports. They composed guides: recall John Conolly's *Working-Man's Companion*. And they produced a host of "penny tracts."

The penny tract is an intriguing tale. Since the closing decades of the eighteenth century tracts of a religious character had been directed chiefly at the poor. These were often short stories that carried a moral. A popular children's writer, Mrs. Sherwood, wrote many such tracts, including one called "The Penny Tract." Written in 1822, it tells the uplifting tale of the impact of a single tract on a wretched household. Here is a taste of this penny fare.

The tale opens with two wealthy sisters sitting in a garden. An old man, with a basket on his arm, walks into the garden. He is selling tracts and when he is asked what they might do with them: "Give them to your poor neighbours, or your servants, lady, if you have no use for them yourself," answered the hawker. The sisters decide to purchase one of the tracts, and shortly thereafter a "miserable ragged woman" came begging. The sisters offered her six pence and the tract. The beggar replied, "I cannot read, miss, more the pity; but my husband is an extraordinary good scholar."[22]

We soon learn what a wretch her "scholarly" husband really is; for though he is an excellent workman, he lacks discipline, tends to drink, and treats his poor wife harshly. She gives him the tract, but he has no interest in it. Later, bored, he takes it up. He tosses it aside after he finishes it: "But, although the poor man could throw away the book, it was well for him that he could not so easily throw away the ideas which the book had just put into his head, though he did his utmost to get rid of them." And sure enough, the tract's message works upon him and leads him to the local minister. "But, not to lengthen out this story too far," writes Mrs. Sherwood, "I must say, in a few words, that it pleased the Almighty that the little tract, sent by the young ladies to this poor man, should be the beginning of a very great change in him; and thus the Lord often blesses a very small thing to the production of some mighty work."[23]

Mrs. Sherwood's tale goes on to chronicle the struggle and success of the poor Frank Downes, and even brings the wealthy sisters back into the tale to wrap things up nicely. It is perhaps not very surprising that such tracts would have been popular in an age of such profound dislocation. An article on the "Literature of the People" in the *Athenaeum* tells us over six hundred thousand tracts were sold from the "Oiled Feather" series alone.[24] What is far more surprising is the new sort of tract that began to appear in 1826.

In 1826 a new venture took shape, the Society for the Diffusion of Useful Knowledge, as opposed to the long-standing Society for the Promotion of Christian Knowledge. It, too, offered penny tracts from a wheelbarrow, but instead of fables and homilies, these were short essays on mechanics or hydrodynamics. Nothing, I think, captures the enthusiasm for the sciences as richly as this effort by the intelligentsia to make systematic discussions of science available to "everyman" at a cheap price. Thirty-two sheets folded and fastened along the fold: sixty-four pages, double-columned in fine print. Here was the proverbial foot up the ladder: essays on chemistry and the new science of electricity, on heat, and on light.[25]

We can capture the character of these penny tracts by turning to the *Penny Cyclopaedia*. This was the Society's boldest project. Beginning in 1833, the Society offered an extraordinary series of tracts that would together culminate in a full-fledged encyclopedia. Here is a sampling from the entry on heat: an entry, I might add, over five thousand words in length, as was Mrs. Sherwood's "The Penny Tract." This excerpt sketches two prevalent conceptions of the nature of heat. The first conceived of heat as a material substance; fine, even ethereal, but nevertheless a form of matter. We might compare this to intuitive notions about the character of electrical charge. The second theory saw heat as a quality of behavior, what the author calls an accident of matter— more simply, a dance. The greater the heat, the more intense the dance. The modern understanding, the kinetic theory, looks at heat in this way.[26]

> As to the nature of heat, whether it should be regarded as a substance or an accident, has been discussed from the time of Bacon to the present day. Those who regard it as having a material existence suppose that a subtle fluid, called calorie, capable of permeating the densest substances, is universally diffused; that its parts are mutually repulsive, but are attracted by the material particles of bodies, and hence they account for the expansions and contractions of bodies, while the effects of radiant heat are explained on principles analogous to those on which the undulatory theory of light is founded.
>
> Those who regard heat as only accidental to matter rest their opinion on the fact that the artificial production of heat is accompanied by vibratory motions in the material molecules of the heated substances. The measure of the quantity of heat produced mechanically would on this hypothesis have a direct connexion with the sum of the vis viva (now called kinetic energy) of the system of vibrating particles.

This is not simple storytelling, and it is hard to believe that it could have been as popular as the tracts of Mrs. Sherwood. It wasn't. The tracts of the Society for the Diffusion of Useful Knowledge were never as successful as those of their religious brethren. Yet their projects were more than gestures by the intelligentsia. There was a genuine interest in learning. We have seen this already in the remarkable discipline of Francis Place both to educate himself and to build a substantial library. Moreover, there were other efforts in this vein and they only serve to underline the extraordinary efforts individuals often made.

In the latter half of the eighteenth century there emerged subscription libraries. For a modest fee, one joined a collective whose funds secured a room and a growing library. You can still go to the Leeds Library, which was founded in 1768. The annual fee was 5 shillings, and this gave you the right to read books at the library, often a room above a shop or at a home. But around the turn into the nineteenth century a variation on this theme was played out: mechanics' institutes. These, too, entitled the member to a reading room and library for a modest subscription. But in addition, they sponsored courses of lectures held in the evenings. Can you imagine working at a bench or tending some large machine of industry for twelve hours and then taking yourself off to attend a lecture on a practical aspect of the sciences, from the workings of machinery to the chemistry of dyes? It is a wonder of the human spirit captured here in a snippet from an address delivered at the opening of a mechanics' institute in June of 1825 at Ashton-under-Lyne: "The desire of knowledge spreads with each effort to satisfy it. The sacred thirst of science is becoming epidemic; and we look forward to the day when the laws of matter and of mind shall be known to all men."[27] (Ashton, by the way, is now part of greater Manchester.) Late in the nineteenth century the *Imperial Gazetteer* observed that before the introduction of the cotton trade and textile manufacturing in 1769 Ashton was considered "bare, wet, and almost worthless."[28] With industrialization it was boom time, and by August of 1819 Ashton sent some two thousand men, women, and children to St. Peter's Field just a few miles down the road out of the sixty to eighty thousand gathered there to listen to Henry Hunt demand parliamentary reform.[29]

Now, six years after Peterloo, the townsfolk had come together to form a mechanics' institute. It is worth noting that the opening address goes out of its way to say that the institute is not a charitable organization and that whatever monies came from the higher classes was offered as encouragement to the working orders. It is also worth noting that mechanics' institutes and the Society for the Diffusion of Useful Knowledge were not without their critics. A Reverend Grinfield produced a pamphlet critical of these efforts which was reviewed in the *Edinburgh Review*. "Let them become conversant with Morals and History and Biography, before we introduce them to Chemistry, Hydrostatics, or Astronomy," the Reverend wrote. The reviewer does not deny the value of such studies, but goes on to assert that "the minds of the working orders are now arriving at such a degree of strength and maturity that they will no longer be satisfied with the simple food which contented their forefathers."[30] The sciences offer a more substantive fare, a fare that is empowering.

There is something here—something beyond the sort of interest we might associate with a David Attenborough special on public television or a Ken Burns series on national parks. There was a draw to science for the working man; it was the way through the thicket.

It is worth doubling back for an episode from the career of Francis Place. Place had been a dedicated supporter of the London Mechanics' Institute, and I cannot help but stop here to note a remarkable exchange between Place and Lord Grosvenor whom Place had approached for a donation. Here is Place's account: "He said he had a strong desire to assist the institution, but he had also some apprehension that the education the people were getting would make them discontented with the government. I said the whole mass of the people *were* discontented with the government, and that although teaching them would not remove their discontent, it would make them less disposed to turbulence."[31] Quintessential Place, is it not? Sadly, but not surprisingly, Lord Grosvenor gave nothing.

Despite Lord Grosvenor, the Institute came into being, and later would become Birkbeck College of the University of London. Place wrote about the first set of lectures of the Institute in March of 1824, observing that eight to nine hundred respectable looking mechanics paid marked attention to a lecture in chemistry.

The British Association

We can appreciate to some extent then how thousands of miners and industrial workers would choose to attend a talk by a university science professor. Yet we should ask as well how Sedgwick came to give a public talk in the first place.

Sedgwick was attending the British Association for the Advancement of Science meeting at Newcastle. The Association's annual meetings were peripatetic, moving from one provincial capital to another. They had become quite the occasion with evening meetings that drew sizeable crowds.

From the beginning, the Association worked to foster a general regard for the sciences. In 1838 the Association was still but a fledgling. It had risen from a gathering in York in 1831, when various members of the scientific community came together in search of a different kind of organization from the typical scientific society. There was nothing wrong, of course, with the Geological Society of London or the Royal Astronomical Society, but there was a sense that something else was needed. They wanted something more inclusive, less elitist; something more like the provincial societies, such as the Manchester Literary and Philosophical Society. The Manchester Lit and Phil had begun in the 1780s. Gentlemen farmers and the landed elite came together every two weeks as a social occasion and as an opportunity to examine a host of matters: everything from fertilizers and crop rotations, or remarks on a recent trip abroad, to experiments in chemistry.[32]

With industrialization the boundaries of traditional life had given way to a greatly expanded horizon. Cottage industries like pin-making had become factories whose furnaces and forges were never quiet. Everything was more complicated, more vested in expertise. The old ways were no longer good enough, and this led to an increased regard for those who cultivated learning and none more so than men of science.

England was now well into the Industrial Revolution and with industrialization had come a burst of enthusiasm for mechanics and the engines of industry. The ultimate icon of this interest would be the Crystal Palace Exhibition of 1851, so called because of its magnificent steel and glass structures. One of these, the Hall of Science, featured towering coal driven machines, several of which became the foundation of the collection for a science museum on Exhibition Row. The new British Association would address a broad audience. A rising middle class was keen on the developments of the day.

The main work of the Association was played out in sessions on an encyclopedic array of topics from fossil fish to ferric oxides. They were led by men of science and open to all members. But the meetings also featured public lectures to broad audiences. Few could do this as effectively as Sedgwick, and when it was announced that he would lead a geological tour at the Newcastle meeting, several hundred rode on river boats from the meeting to the beach that morning, looking forward to his talk—not to mention the thousands of working men who showed up on their own.

The Event

All this, then, gives a context for that remarkable event at the beach outside Newcastle. In a letter to his niece a few days later, Sedgwick writes:

> On the Friday of the Association week, I went to the mouth of the Tyne with a geological class of several hundreds, and nearly all the population of Tynemouth turned out to join us. You would have been amused at the picturesque group, clustering among the rugged precipices of a noble sea-cliff or congregating on the sand below, while I addressed them at the utmost stretch of my voice six different times, from some projecting ledge that served as a natural pulpit. Every one was in high spirits, the day was glorious, and we all returned up the river in steam-boats so as to join in the work of the evening meetings.[33]

Such was the standing of science and its ministry.

A summer day at Tyneside, warm and sunny, but curiously still; unlike the mornings and evenings when the beach is swept by strong breezes that fail to carry the promise of summer. Waves then slap the shore, unlike the quiet lapping of mid-afternoon. We can imagine a worker down on the beach who smells his long-forgotten childhood in the stillness so far removed from the incessant din of steam engines and the whine of conveyor belts. He can feel his youth in the glint of the sun, neither busy nor unruly, but simply absent from his long days inside. The professor has not started yet. You can see him talking with others, catch the way he leans forward with attentiveness and speaks animatedly. He seems a kindly man as he bends closer to his friends. As the others leave the professor alone and he steps onto a rock, you slip your boots off and push your feet into the sand, wondering if you will be able to follow him. It would be a shame if he leaves you behind.

But the good professor did not leave him behind. We may well wonder what Sedgwick could offer this "picturesque group" that so captivated thousands of miners and factory hands, as well as gentle ladies and their escorts. What held them spellbound at the meaning of rock, ore, and sea, science and commerce, hard work and contemplation?

We can rest assured it would not have been a simple recital of geological fact or fancy. It would not have been a dissertation, dry as dust, on the recent controversy over the rocks of Devon or the difficulties of working out the boundary between the Cambrian and Silurian systems. But nor would it have been mere platitudes on the noble effort to understand the history of the earth. Instead Sedgwick presented a grand vision, a vision of ancient life trapped in a mudstone eons ago that now, in a steam-driven economy, is worth a king's ransom, not because like gold or diamonds it may adorn the wealthy, but because it fuels an entire society: giving good work to the many who produce an abundance of goods. And so, we rightly stand in awe of the connectedness of nature's many parts over space and time—as our god-given capacity of reason transforms the spent life of ancient ferns into the fuel that drives our industry and our society.

In an age of transition, Sedgwick spoke from a rock.

THE IDEA WITHIN SYSTEMS

On December 17, 1832, just three days after the newly reformed parliament had been elected, the election that saw George Grote become an MP, the Master, fellows, and students of Trinity College met in the chapel for an annual service to honor the traditions of the college and remember the great scholars of its past. At the pulpit was the Rev. Adam Sedgwick, Woodwardian Professor of Geology. Sedgwick took advantage of the occasion to reflect on the character of a Christian education in this new era. His sermon offers us a glimpse of what he would later share with the thousands gathered on the beaches of Newcastle; here is the very heart of Victorian science.

Recall once again Bulwer-Lytton's remark that transformation had become visible for this generation. Sedgwick felt the need to re-examine the college's practice and to take its measure. What was the purpose of a university education? His sermon so resonated with its audience that he was approached by his friend and colleague, William Whewell: would he consider publishing his remarks? Flattered, Sedgwick consented and so we have *A Discourse on the Studies of the University*, a work that resonated with a wider audience as well. It went through five editions. And by the fifth edition in 1850, Sedgwick had added 442 pages of preface and over two hundred pages of appendices. As the central essay remained unchanged, Sedgwick would remark it had become "a grain of wheat between two millstones."

It is intriguing to note that the original *Discourse* was reprinted in 1969, another time when transformation was visible.

Sedgwick was a master scientist and a scholar of considerable range, but what we see here most especially is his gift as a teacher. His examination of a Christian education was not a matter of ritual pieties. He frames the call for the right sort of education as directly for his young audience as one could, offering the observation: "What a melancholy contrast we too often find between the generous temper of youth, and the cold calculating spirit of a later period!" What was called for was an education to help you grow into the man you would want to become, that would foster "habits of practical kindness, and self-control" and cultivate "all those qualities which give elevation to the moral and intellectual character."[34] A message equally apt for 1832, for 1969, and for our own day.

Sedgwick then proceeded to examine the studies at the university from this perspective, beginning with the sciences and focusing on the relationship between science and religion, a matter that was a constant element in the background of geology across the early nineteenth century.

AN ASIDE ON GEOLOGY AS IT STOOD

On becoming the Woodwardian Professor in 1818, Sedgwick threw himself into geology. The prevailing framework of the day had been laid out by the threefold master work of Georges Cuvier, but before we take up Cuvier's work, let's step back a bit to consider what geology is about.

Look across a landscape. The sciences can tell us lots about what is there, about the plants and animals, about cells and the array of processes within living things. And we can learn about the composition of everything out there, what things are made of and how they get made, and then there's the water cycle, erosion, and the host of other physical processes at play in what we see. But how can we come to see what it all was like a hundred years ago, or a million? How old are the mountains, and is everything equally old? If not, then where and how are new rocks and mountains made?

Most efforts to make sense of such questions had been guided by a biblical framework: the Earth was seen as not especially old, and essentially the same now as when Adam and Eve stepped out from the Garden of Eden. The only genuine transforming event had been Noah's flood. But toward the end of the eighteenth century things became quite charged, with intriguing arguments about a vastly older Earth and conjectures that stretched the boundaries of the biblical account.

Now we can take up Cuvier's formative work. There was first his remarkable ability to bring to life the fragmented forms of ancient animals. If it is patently clear to us, as we glance at museum cases or crane our necks to take in the skeletal models of huge lumbering beasts, that these rocks bear unmistakable signs of former life, we must remind ourselves of the profoundly fragmentary and unfamiliar face they most often present. The pages of any book of fossils present a myriad of broken designs, swirls of line and contour that may, at times, speak clearly of former life, but certainly not all the time. It took an eye well trained in the lines and processes of life to trace the face of an organic past in these broken rocks.

It had been decided before Cuvier's day that fossils were monuments of ancient life, but it was his genius that gave them that life. He was able to do so because, in Sherlock Holmes-ian fashion, he could make a fragment speak to the whole. As Holmes could see the health, history, and habits of a man in his hat, so could Cuvier see the corresponding history and habits of an ancient beast in fossilized teeth and bits of bone.

The second fold of Cuvier's master work also involved fossils, but to a different end. Working with Alexandre Brongniart, Cuvier effectively showed how fossils could be used to identify beds of rock and so solve the problem of correlation.

Correlating the rocks of one locale with those of another is problematic for several reasons. Beds of rock are not uniform. They can thin out and disappear. They also are not flat; that is, they are often variously bent and contorted. Finally, there is a limited variety of physical characteristics: sedimentary rocks, mudstones, sandstones, and limestones, for example, come in a limited variety of colors and textures. Thus it is hard to know if a bed of sandstone on your estate is the same as a bed several miles down the way. I say "sandstone" for a reason. The Industrial Revolution had made coal, already a useful commodity, a most valuable one, and sandstone played an intriguing role in coal exploration.

It turns out coal beds were always sandwiched between sandstones above and sandstones below. If you owned property and found a sandstone bed, there was a possibility that there was a king's ransom worth of coal in the ground below. But you needed to know if that bed of sandstone was of the type found above coal, called the New Red or if it was the Old Red below. To not know was maddening, because digging mines was very costly, and many a landowner went broke vainly seeking coal below his sandstones.

Cuvier and Brongniart solved this problem by looking at the fossils within beds of rock; for the fossils of the New Red above differed from those of the Old Red below. While fossils had long been collected and treasured as curiosities, no one had paid the least regard to the order and regularity with which Nature had placed them. But now fossils were no longer *idle* curiosities. They were markers, signs. They had been carefully disposed by Nature; each assigned its proper place. By identifying strata, fossils enabled geologists to work out the relative ages of these beds of rock, with younger rock resting atop older rock. This, in turn, enabled them to visualize the layers of strata beneath the surface and to create stratigraphical maps for entire regions.[35]

There was yet a third commanding element to Cuvier's work in geology: the notion of a geological revolution which structured a powerful theory about the overall sweep of the earth's history. Writing in 1812, Cuvier envisioned the earth's history as having been scarred by an array of catastrophes of a magnitude beyond the scale of nature as we know it. We seem to stand before a deep chasm as we read: "The thread of operation is here broken, the march of nature is changed, and none of the agents she now employs were sufficient for the production of her ancient works."[36] For Cuvier, as for Pindar, the past is only half-brother to the present.

What had fired Cuvier's imagination was the problem of extinction. He had brought the fossil record to life using the anatomical rules and regularities of living animals. Why, then, were these fossilized forms no longer alive? Earlier understandings of fossils had seen them as marginalized, failed forms. But now it was clear they had been perfectly viable. Why then had they become extinct?

Because of revolutions.

Cuvier envisioned great mountain-building episodes, where powerful forces deep within the earth were unleashed, raising sea beds into the sky and transforming ocean depths into mountain peaks. Ensuing tidal waves of displaced ocean waters would wash across the landscapes of the day, burying all living things. It would not matter how viable an animal had been at grazing or hunting in the face of such violence.

This was an exceptional theory, solving the problem of extinction and explaining such striking phenomena as finding fossils of ancient sea life on the uppermost reaches of mountains, such as the Alps. Further, it laid before the geological community well-defined problems like the basic enterprise of connecting revolutions evident in overturned strata and breaks in the fossil record in one locale with similar phenomena in other regions.

Cuvier was particularly drawn to the last great revolution. In a chapter, titled "Proofs, from Traditions, of a great Catastrophe," he hypothesized that a small remnant of humanity had survived the last revolution. He then sought the connection between it and the records in ancient storytelling of a great flood. There were flood epics not only in the Old Testament, but in epics from other ancient societies, from Babylon, Greece, India, China, and elsewhere. The most likely haven for the survivors of a worldwide flood would have been the Himalayas. Not only were these the highest valleys on earth, but William Jones had recently argued that you could trace the history of languages back to a mother tongue, Sanskrit, the holy language of ancient India. In Cuvier's view, the survivors had grown and multiplied, migrated across the globe and in the many new languages that developed over time with these wanderings, the story of an ancient flood that had virtually destroyed all life on earth was told and retold.

Such was the context for the practice of geology when Sedgwick became Woodwardian Professor in 1818. It was a context, we may add, coherent with Mitford's approach to history. For both Cuvier and Mitford, the aim of historical scholarship was a concordance of events, be they colonial plantings or geological revolutions, and the construction of an overall chronological sequence.

SEDGWICK CHALLENGES CUVIER'S GEOLOGY

So in 1818 Sedgwick threw himself into the new stratigraphical geology and was so successful that he was elected president of the Geological Society of London in 1829. By that time Sedgwick was in a different place. He had questioned both geological revolutions and the idea that the sciences and religion shared a common language.

On the evening of February 19, 1830, the fellows of the Geological Society met on the occasion of the Society's twenty-third anniversary. It was the custom to begin these evenings with a meal at the Crown and Anchor, before adjourning to the Society's chambers not far away in Somerset House, where the president would speak to the state of geology and its progress. It is left to the imagination whether the members had to work their way through a winter night like the one Dickens here describes, as they walked from the Crown and Anchor:

> London . . . As much mud in the streets, as if the waters had but newly retired from the face of the earth, and it would not be wonderful to meet a Megalosaurus, forty feet long or so, waddling like an elephantine lizard up Holborn Hill. Smoke lowering from the chimney-pots, making a soft black drizzle, with flakes of soot in it as big as full-grown snow flakes—gone into mourning, one might imagine, for the death of the sun.[37]

In any case, Megalosaurus or no, let us take up Sedgwick's address with an eye on the relationship between science and religion. Toward the close of his address, reflecting on the character of geology, Sedgwick sharply rebuked scriptural or mosaical geology. His comments were provoked by a recent publication by a fellow of the Society,

Dr. Andrew Ure, a work incompetent in detail and perverse in principle. It had failed utterly to get its facts straight, and had drawn its generalizations from the biblical and not the geological record. This was a categorical error; for "to seek for an exposition of the phaenomena of the natural world among the records of the moral destinies of mankind, would be as unwise, as to look for rules of moral government among the laws of chemical combinations."[38]

Sedgwick would return to these matters in his presidential address the following year. This time his remarks were touched off not by the marginal hypotheses of a Scriptural geologist, but by a central hypothesis of the Cuvierian program. Sedgwick urged that the signs of flooding scattered over much of the world "do not belong to one violent and transitory period."[39] They are not evidence of a single grand revolution. Sedgwick reviewed various papers, but there was more to it than the facts of field and fossil. To deny Cuvier's hypothesis was not to deny Genesis. Sedgwick explained that while moral and physical truth may have a common essence, their foundations rested upon two very different types of bedrock: "And in the narrations of a great fatal catastrophe, handed down to us, not in our sacred books only, but in the traditions of all nations, there is not a word to justify us in looking to any mere physical monuments as the intelligible records of that event."[40] Noah's deluge was a moral, not a physical event.

In an essay on Coleridge, John Stuart Mill agrees with him that across much of Christianity's history bibliolatry, as in idolatry, had prevailed; a superstitious worship of texts that had led to the persecution of Galileo and in his own day led many to see the discoveries of geology as anathema.[41] Sedgwick was also critical of bibliolatry: the purpose of learning science was not to establish the authority of scriptures. It was, instead, to establish natural religion—that religion whose sacred text is nature.

This notion, that there are two sacred texts, the revealed word and the natural world, reached deeply across Christian thought, and at its core lays the argument from design. If there is design, then there was a designer: an argument witnessed when we would look at the flotsam and jetsam of a beach arrayed to spell out "Johnny loves Mary" and infer that Johnny, Mary, or some other soul had pulled the seaweed, egg cases, and whatever else to spell out the message. The play of the waves is no match for such designfulness. As there are myriads of designful arrays in nature, these must have been brought together by some agent outside the random concourse of atoms, an agent otherwise known as God.

It is clear that there is something about designfulness itself that requires a designer. At the same time, we know that nature is full of patterns and designs. We can see them everywhere. How many coffee-table books have there been, filled with extraordinary images of everything from the intricacies of snowflakes to the birth of stars in distant nebulae? Science is awash in design. This is what scientific laws and concepts are all about. This is what they explain.

So we have a curious tension. The argument from design takes off from the notion that there is designfulness outside of what "naturally" occurs, while what naturally occurs is itself saturated in designfulness. It is no wonder that issues surrounding design can sometimes be full of sound and fury, and signify far too little. Lines have been drawn that have put science and design in opposition, where opposition is hardly the case.

We may now go back to Sedgwick's *Discourse*, delivered over a year after the second of his presidential addresses, and better appreciate his remarks on science and religion, pausing only to observe that so much of contemporary discussions of the purpose of an education and most especially of a scientific education is wrapped around employment "pipelines." Not so Sedgwick's discourse. The purpose of scientific study at the university is "to see the finger of God in all things animate and inanimate," and in following natural law to "find it terminating in beauty, and harmony, and order."[42]

BACK TO SEDGWICK'S SERMON

In the chapel at Trinity College Sedgwick offers two arguments from design. The first addresses the imagination. He notes that music has no charms for the deaf, nor painting for the blind. If our senses were not preordained to be in sync with the world around us, then we would be deaf and blind to the world. That is, in the act of turning to experience we witness the hand of God in the very fact that we can witness phenomena. The other argument comes from the harmony of the interworking parts of living things. As he puts it: "Contrivance proves design: in every organic being we survey (and how countless are the forms and functions of such beings!) we see a new instance of contrivance and a new manifestation of intelligent superintending power."[43] This arena for design was a central aspect of biology and sat at the core of the reconstruction of fossil fragments.

Fair enough. Sedgwick has linked the purpose of scientific study to an appreciation of natural religion. But there is more, and this takes us back to Sedgwick's claim that Noah's flood was a moral event.

In the *Discourse*, Sedgwick repeats his sentiments from his presidential address: "But if the Bible be a rule of life and faith—a record of our moral destinies—it is not (I repeat), nor does it pretend to be, a revelation of natural science." But he takes it further. "These writings deal not in logical distinctions or rigid definitions. They were addressed to the heart and understanding, in popular forms of speech such as men could readily comprehend." He then illustrates his thesis with an example that has nothing to do with the sciences: "When they describe the Almighty as a being capable of jealousy, love, anger, repentance, and other like passions, they use a language accommodated to our wants and capacities, and God is put before us in the semblance of humanity."[44]

This approach to the Bible—the principle of accommodation—has a history reaching at least as far back as Galileo. As we noted earlier, the Copernican hypothesis had been challenged on the ground that it conflicted with passages in the Bible, such as the passage where Joshua prays that the sun stand still in the sky and his prayer is met. If Copernicus were right, Joshua should have prayed that the earth cease spinning on its axis. But Galileo disagreed. While the Bible does not speak untruths, it is often the case that the true meaning of a passage is obscure and does not lie in the "unadorned grammatical meaning" as it presents itself; adding: "Thus it would be necessary to assign to God feet, hands and eyes, as well as corporeal and human affections, such as anger, repentance, hatred, and sometimes even the forgetting of things past and ignorance of those to come."[45] This is very much in the spirit of Sedgwick's comments some two hundred years later.

Galileo goes on to suggest: "These propositions uttered by the Holy Ghost were set down in that manner by the sacred scribes in order to accommodate them to the capacities of the common people, who are rude and unlearned."[46] A battlefield was not the time for God to pause and explain to Joshua that he should be praying for the earth's spin to cease, not for the sun to stand still. The point of this passage was not astronomy, but the relationship between the ancient Hebrews and the Almighty.

The principle of accommodation sets bibliolatry aside. It is not interested in the literal import of the Word. The language of the Bible, be it simple and direct, meta-phorical, or symbolic, could not be read directly. What the Word meant could only be approached through an understanding of the spirit of that age. We have seen the heart of this approach to interpretation in Thirlwall's prefatory remarks to the Schleiermacher essay and in his *History*. One needs to read the ancient tale in terms of the understanding of the ancient storyteller. Sound history lays in an appreciation of the genius of an age.

This same appreciation of the genius of an age will be found at the heart of Sedgwick's new geology of systems.

SEDGWICK'S NEW GEOLOGY

These were heady days for Sedgwick, starting with his fieldwork in the Eastern Alps in the summer of 1829, reaching across his two years as president of the Geological Society, and his renewed focus on the older rocks of Wales. We must add to these not only his professorial responsibilities for lectures and fieldwork with his students, but also his role as a leading liberal within the Cambridge community and the demands on his time for parliamentary elections in both 1831 and 1832. Thus we find him writing to his colleague Roderick Murchison late in the summer of 1831, complaining that he'd only been able to break away for fieldwork the first of August. As it happened, he took along with him an undergraduate who was keen on improving his geology. Many years later, Darwin would write fondly of this trek with Sedgwick; how Sedgwick taught him to trust his own judgment by having him walk in parallel at some distance off, after which they would come together and compare notes on what they had seen and how it might be interpreted.[47]

Sedgwick continued in Wales after Darwin's two weeks with him, deciding not to go to the inaugural gathering of the British Association in York that September. He would make up for that the following year by attending the Oxford meeting in June of 1832, and then agreeing to preside at the third meeting, at Cambridge, in 1833. He had more summer left after Oxford, so he went back to Wales, returning to Cambridge for the start of the Michaelmas term. Not long after was another parliamentary election, the one where Grote was returned for the first reformed Parliament, and but a week later on Monday, December 17, he gave the sermon that became the *Discourse*.

But it was not just these professional duties that made these days so intense. Sedgwick had turned a corner in his work, a turn that led to the concept of geological systems. Any analytical discipline is deeply indebted to its past, and this is certainly true in the historical sciences. A sound geological argument is a complex entity. The issues surrounding its character are rich and varied, and they have emerged from a complex history which has everywhere shaped what problems are the most pressing and what frameworks the most promising.

Systems have played a central role in structuring the earth's history since they were first developed in the 1830s. There was a coherence to this work which transformed ill-defined strata into a succession of well-formed natural wholes. This broad coherence is our goal. What did systems mean? What problems did they solve? And what connections did they forge?

For Cuvierian geology the units of history were delineated by discrete events, catastrophic revolutions. Between each set of violent endpoints lay a continuum of repose where nature had been as quiet as we see in it the modern era. A system was just the opposite. It was not marked by its endpoints, but rather by the character of the segment itself. The new systems geology defined the chapters of the earth's history by the character of the organic remains found within the formations, the term "systems" reflecting the importance of whole assemblages of life forms. Decades and centuries are units of time marked by their endpoints, but when we think of the Renaissance or the Industrial Revolution the endpoints are up for grabs. They are defined, instead, by deep changes reflected in the character of their middles. Like a cloud they do not have doors that mark passages, but still you know when you are within one.

The movement from the Cuvierian program to systems played out in several steps, but it was clear there was something new going on. William Conybeare caught it in his report to the British Association on the recent history of geological studies at the Oxford meeting in 1832.[48]

And then there was Phillips's introductory chapter.

Topographical Geology

The first volume of John Phillips's *Illustrations of the Geology of Yorkshire* appeared in 1829. With the release of the second volume in 1835 volume one was reissued, but with a striking alteration: the whole of the original introductory chapter was gone. In the new preface, Phillips simply noted that the chapter was no longer appropriate to a "topographical work suited to the year 1835."[49]

What had changed so between 1829 and 1835? What had Phillips come to see as so out of date that it required not a simple modification or more substantial revision, but a complete erasure? And what is a topographical work? The answers take us straight to Sedgwick and the new geology of systems.

Phillips's 1829 introductory chapter had not been simple matters of fact which further research had displaced. It laid out the general principles of geology as a framework for the analysis of Yorkshire that would follow, principles we traced in our discussion of the Cuvierian program: a universal geology aimed at mapping the earth's strata, which "may be discriminated and identified by their organic contents" and giving place to powerful revolutions, such as the Deluge.[50]

Phillips, by the way, attributed the use of distinctive fossils to identify strata to William Smith, who had made this discovery about the same time and independently of Cuvier, using it to develop the first stratigraphical maps of England.

Volume one had been dedicated to Smith. In its place now stood "topographical" geology. Volume two was dedicated to the Reverend Adam Sedgwick.

The Eastern Alps Thesis

Sedgwick and Murchison made plans to tour the Austrian Alps in the summer of 1829. Sedgwick rushed about, presiding over the last meetings of the Geological Society, writing up several papers he had read to the Society that year, and there was university politics (recall that delightful episode with the Paddington coach packed with Macaulay and friends) but he was able to meet the boat, and the tour proved to be a great success.

Their initial core discovery was the absence of any break between secondary and tertiary rocks in the Eastern Alps strata, a break prominent elsewhere. As Sedgwick and Murchison put it, this contradicted the geological record in England and France where these formations are "discordant and they have not perhaps a single fossil in common."[51] This boundary had been the only boundary where there was absolutely no evidence of a gradual passage. Their finding was the straw that broke the Cuvierian revolution's back. A new set of rules was going to be needed. Here was a divide long thought to mark dramatic events separating two of the most fundamental chapters in the history of life, and yet the Alps, a region that had been subject to extraordinary forces of upheaval both before and after, was quiet and division was replaced by conformable passage in both the physical and fossil evidence.

This was not to deny that there had ever been violent events, but they were not key stratigraphical markers. They had a local significance, and that was all. A regional geology was thus the first new rule, and we can see it emphasized by Sedgwick in his presidential address the following February. For example, in his review of significant papers presented to the Society that year, Sedgwick highlighted a paper by Conybeare that urged that the Thames River valley had been carved by a massive flood and a paper by Lyell and Murchison that argued that valleys in the Auvernge region of France had been carved without such violent washes of water. Sedgwick was not trying to appease both sides of a controversy, but instead to affirm that what was true of England need not have been true elsewhere.

The second new rule addressed Cuvier's and Smith's notion that strata could be identified by distinct fossils remains and so linked across separated regions. Sedgwick and Murchison became convinced that there was a general correspondence between several Alpine and British formations, even though there were no specific fossil identities.[52] It is not that they had watered down Cuvier's notion of specific identities into a weaker tea of mere similarity. Rather, specific identities failed to carry the weight of the natural relationship between these widely separated regions. This new approach accommodated differences that were the natural result of the integrity of distinct regions, and yet was sensitive to an overall resemblance—as we might think of the broad similarities in Spanish and French despite many particular differences.

Moreover, if you batched the fossils by the kind of embedding rock—mudstone, sandstone, or limestone—then the similarities were heightened. Sedgwick saw the embedding rocks as clues to ancient habitats. Just as some living animals require a muddy shore, so fossils of kindred species occur abundantly in shales, while other animals, such as corals, build calcareous deposits and their fossil kind are found in limestones. They were reconstructing entire ecosystems of the distant past.

If we go back to Thirlwall's approach to ancient heroes (Hercules, Theseus, etc.), we see a basic conformity with this new rule. Thirlwall saw myth as "alive" in that it responded to the needs of developing communities. Hercules the greater becomes Hercules the lesser as the issues of everyday life move from basic questions about community to matters of propriety within the more complex network of a society. If we let Hercules and Theseus represent different locales, then the correspondence in the changes of these tales, from greater to lesser, is showing how these legends are adapting to new conditions. The differences between wrestling the Nemean lion and slaying the Minotaur are of little consequence. There is an integrity to the distinct deposits that allows for specific differences. But the overall resemblance is what is telling, for it points to common underlying forces.

As our study of 1832 progresses we find ourselves tripping over an array of connections. Some are mere happenstance, as in Sedgwick busily working in the election when Grote would win his seat in Parliament, but others are far richer: the strong similarities in the approach to history in the work of Cuvier and Mitford, and the corresponding similarities between Thirlwall's approach to myths and Sedgwick's study of fossils. Hurrell Froude's remarks on our Trinity College scholars come to mind. That sense that they fed off of one another, catching resonances as Whewell's book, or Sedgwick's lectures, Thirlwall's research, or Hare's taste "pop upon one at every turn." The historical meaning of both fossil and myth shifted from discrete particulars to broader underlying resonances. That is, Sedgwick and Thirlwall each saw powerful parallels between developments in different locales as grounds for postulating an underlying commonality, despite incidental differences. Our study is pursuing the same end. There is an overall conformity to Sedgwick's and Thirlwall's analyses, and though fossil and myth are decidedly different, the conformity speaks to a common underlying analytical structure that has shaped their work.

A Horizontal Gaze

There had been a vertical push to the Cuvierian program. The central task was to sort strata by their distinctive fossils and then work out a vertical or chronological sequence. Picture the hundreds of layers of the Grand Canyon. The older rocks are at the bottom. If you can link strata anywhere to one of those layers, then everything fits into a coordinated series.

But with the Eastern Alps thesis one's gaze moves horizontally instead. By following a single layer you are moving along a "moment" in time, a geological moment being however long it took to collect that layer. This profoundly changes the complexity of things. In his presidential address of 1830 Sedgwick offered an example: suppose you examine two strata, one atop the other. They contain strikingly distinct fossils. But if you follow them horizontally, you find that they "open up." Other beds of rock intervene and before long you uncover formations of considerable thickness altogether. And so we are "no longer startled at the change of organic types in the west of England, between the coal measures and the lias. For between the times of their deposition, there were completed at least five great geological periods."[53] A seeming discontinuity, a jarring break in the series of life forms suggestive of a revolution, is translated into a series of intervening steps and a vast span of time.

After the tour to the Austrian Alps, Sedgwick continued to work at a regional geology which follows strata horizontally to reconstruct ancient landscapes. The complexity of this work is stunning, whether we look at the Lake District, at Wales, or his important work with Murchison in Devon.[54] But there's a nicely illustrative application of Sedgwick's horizontal gaze in Phillips's study of Yorkshire in volume two.

Phillips's second volume was given to the mountain limestone district, a large formation resting atop the Old Red Sandstone. The crux of his analysis concerned the differences within this formation as one moved from the southeast to the northwest. The southern section was of a generally uniform and simple character, while the northern series had many layers, with distinct beds of limestone separated by bands of sandstone and thin seams of coal—all in a complex and variegated structure. Phillips saw the northern series as "the variable effects of inundations from the land, and the inconstant movements near the shores of the sea," while the southern series reflected "depositions beneath deeper and more tranquil waters."[55]

Phillips's horizontal gaze had led him to evoke an ancient landscape and the local forces which had generated the remains that now characterized this district. Central to this work was the perception that contemporary deposits could differ in their character and that such differences were important signs of ancient environments. For Cuvier, two deposits were contemporary if they held identical fossils. The new geology coming out of the Eastern Alps thesis stood this on its head. Deposits with altogether different fossils could represent different environments from the same historical moment. Cuvier's geology had replaced the landscape with a stratigraphical table and a map derived from discrete identities. The emphasis was on lines of connection, not on the character of the locale itself, and it fostered an increasingly fine-grained set of distinctions. The broader continuities of nature had been sacrificed at the altar of a too-narrow precision. We might paraphrase Sedgwick's praise for Thirlwall's new approach to classical studies. Substance had been overlooked. We have been too taken with "the measure, the garb, and fashion of ancient song, without looking to its living soul or feeling its inspiration."[56] That is precisely what Sedgwick had now done with the fossil record. He had sought its living soul, the many ecosystems of the distant past.

We see the echo here of Thirlwall's break from Mitfordian history. Mitford's approach, with its emphasis on begats to frame a universal chronological sequence within which were placed city foundings, cultural inventions, battles, and migrations, like Cuvier's geology, generated tables of discrete lines of events. For Thirlwall the aim was not to parse history into a myriad of discrete items, but to uncover deep sensibilities that would bind an era—its living soul.

Why Topographical?

Phillips called Sedgwick's new geology "topographical," reminding us that a landscape is equally fit for study by both geologist and painter. In their studies of Victorian art and criticism, both Henry Ladd and Patricia Ball discuss notions of topographical detail that shed light on our study.

In the late eighteenth century an interest in travel literature had generated a group of artisans known as "topographers." Their attention to detail and authentic portrayal was not appreciated by critics of their own day. Fuseli, for example, long a professor at the Royal Academy, scorned their "delineations of a given spot," calling them "mere views" and "map work." Yet they flourished and became the teachers of the next generation, of Constable and of Turner; where the same regard for the authenticity of objects and of place—of *topos*—is found. "However brilliant their poetical effects of light, atmosphere and storm clouds." Henry Ladd underlines, "there was always the understanding on the part of their public that a picture was a picture of a certain place and, even more important, that this was the way that place or that tree or that castle looked."[57]

This is precisely the cardinal principle of the great critic John Ruskin, for whom fidelity to detail was necessary for both animate and inanimate nature: "It is just as impossible to generalize granite and slate, as it is to generalize man or a cow." There is, he wrote, "a science of the aspects of things, as well as of their nature."[58] Authenticity of place, a science of aspects: these are critical notions, close to the most fundamental grounds of perception, analysis, and practice. There had been facts and details before in Cuvier's geology, in Mitford's history, and in Claude's landscapes, but we are seeing a shift here, a different set of criteria for a satisfying rendering.

Susan Cannon talks about this same shift in one of her essays in *Science and Culture*. She notes a distinctive concern for massive amounts of factual material, adding that "an interest in the detailed complexity of the actual may seem to suggest the influence of Romanticism." However, she continues, since Romanticism is neither a massive body nor a force, it could not be an historical agent. We agree. We should turn things around and see Romanticism as the consequence of a certain regard for the complexity of the actual (the fourteen-sidedness of things). That is, we need to see how within broad intellectual movements like Romanticism there lies a set of perceptions on the nature, structure, and form of arguments and their legitimate ends that do the work, that carry the weight of inferences, that bring things to a natural conclusion.[59] This is the "force" of an analytical framework.

Sedgwick's new geology led Phillips to drop his original introductory chapter because things had changed at the most fundamental level of how historical inferences work and the proper ends of analysis. Sedgwick changed what geologists looked for, what they saw, and what it meant.

Every discipline offers a technical language that shapes the material into problems that can be solved; for someone with a hammer, everything begins to look like a nail. Stratigraphical geology had looked through landscapes to see columns of strata identified by fossil markers. These markers, in turn, allowed a given column to be linked to other columns, ultimately finding place within a single grand series representing the entirety of the earth's history. The Eastern Alps thesis is a new tool. It leads to contours rather than columns, to ecological communities within these contours rather than fossil markers, and to the evocation of a regional geology rather than a universal one. This involved changes in practice; for example, early in the 1830s Sedgwick's notebooks become filled with a much more thorough and exact recording of fossil finds—both their location and the character of the embedding rock, where earlier he had been satisfied with a broad sense for locale.[60] And it led to changes in language. As geologists sought to capture the distinctive quality of a given face of rock that set it apart from others of the same epoch or from others of the same broad type, we find such terms as "aspect" and "facies."[61]

There had been a restructuring: a restructuring that parallels the shift we traced in Thirlwall's *History of Greece*. Thirlwall rejected the stratigraphical project of Mitford's approach, denying the value of correlating the several histories of antiquity into a unified march of events. His was a regional history, a history of Greece founded upon the Greek character and the particulars of the Grecian experience. He did not account for the emergence of civility in Greece by tracing connections to colonies from Egypt or Phoenicia, or the invasion of Hellen and his sons, as had the ancient themselves and Mitford with them. He saw these as conceptual devices, like the mighty arm of Hercules. They reflected the early historical imagination not the events of history. The more fundamental matter was the character of the ancient Greek. That was the starting point.

Sedgwick's new geology was also regional, shifting its analysis of fossils from the correlation of deposits to the new project of reconstructing an ancient landscape. Guided by the contextual heuristic of the Eastern Alps thesis, fossils were read in terms of the environment which had brought them forth and the spirit of their age. The aim of history was not the tabular display of relations and the measured march of violent events, but the evocation of the past. To this end it was necessary to examine in close detail the character of a region, a particular place—*topos*. Authenticity became essential, a regard for the local face of things. Systems and topographical geology sought the ancient surface. Thirlwall sought the landscape which was the mythological mind. They both shared that same regard for authenticity, along with brilliant effects of light, atmosphere, and occasional splashes of red that was the genius of Turner.

AT THE BEACH OR AT CHAPEL

Earlier we asked what it was, when Sedgwick bound from rock to boulder on the shore at Tynemouth, that held thousands of miners and factory hands, as well as gentle ladies and their escorts, spellbound at the meaning of rock and sea, science and commerce, hard work and contemplation. What had so deeply captured that audience?

It would not have been a simple recital of geological fact or fancy. Nor would it have been a dissertation, dry as dust, on some geological controversy. There was a quality to the good professor we have glimpsed for ourselves: a capacity to see what things really mean, what issues are really asking of us, and he could give shape to these matters with a voice that was plain and direct, but also stirring. In the chapel at Trinity College he transformed a perfunctory recital of the glories of the great scholars of Trinity's past, the likes of Newton and Milton, by remarking: "What a melancholy contrast we too often find between the generous temper of youth, and the cold calculating spirit of a later period!" The right education was not a matter of learning more, not a matter of becoming more rigorous, not a matter of more science or business or engineering. It was, instead, "to lay a good foundation against the coming time, by fostering habits of practical kindness, and self-control—by mental discipline and study—by cultivating all those qualities which give elevation to the moral and intellectual character"; "in one word, by not wavering between right and wrong, but by learning the great lesson of acting strenuously and unhesitatingly on the light of conscience."[62]

More so than Herschel's study of the heavens, or Faraday's invention of the dynamo, Sedgwick—whether at the beach at Tyneside, from the president's chair at Somerset House, or within Wren's magnificent chapel at Trinity College, with hammer in hand or a pen—captured the sense of his age and its conscience.

NOTES

1. Cannon, *Science in Culture*; Fox, *Caloric Theory of Gases*; Kargon, *Science in Victorian Manchester*.
2. Holmes, *Age of Wonder*; Cannon, *Science in Culture*; Paradis and Postlewait, *Victorian Science and Victorian Values*.
3. Todhunter, *William Whewell, D. D.*, vol. ii, 62.
4. Storella, *"O, what a World of Profit and Delight,"* 89–90.
5. Playfair, *Illustrations of the Huttonian Theory of the Earth*, 8.
6. Herschel, *A Preliminary Discourse on the Study of Natural Philosophy*, 21–22.; see also Hall, "Letter from Captain Basil Hall," 211.
7. Carlyle, *Sartor Resartus*, 42–43.
8. Carlyle, *On Heroes, Hero-Worship and the Heroic in History*, 7.
9. Herschel, *A Preliminary Discourse*, 13–14, 95.
10. Faraday, *Experimental Researches in Electricity*. There are several fine works on Faraday; see Williams, *Michael Faraday*, and Agassi, *Faraday as a Natural Philosopher*.
11. Duhem, *Aim and Structure of Physical Theory*, 70–71.
12. Tyndall, *Faraday as a Discoverer*, 6–8.
13. Lamb, "The Superannuated Man," 122–23.
14. Sedgwick, *Life and Letters*, vol. i, 515–16.
15. Hammond and Hammond, *Lord Shaftesbury*, 16–17.
16. *Ibid.*, 67–69.
17. Popper, "On the Sources," *Conjectures*.

18. "Children's Employment Commission, First Report," testimony of Sarah Gooder, aged eight years, *Parliamentary Papers*, 1842, vols. xv–xvii.

19. Hammond and Hammond, *Lord Shaftesbury*, 72.

20. *Ibid.*, 78.

21. Schweber, "Scientists as Intellectuals," 4–5. Census data began in 1801; see also A Vision of Britain Through Time.

22. Sherwood, "The Penny Tract," vol. xiii, 350–54. This effort to reach the masses of working poor parallels Cobbett's "Political Register"; see Thomson's *Making of the English Working Class*.

23. Sherwood, "Penny Tract," vol. xiii, 354.

24. Dilke, "Literature of the People," 13.

25. Storella, *"O, what a World of Profit and Delight."*

26. *Penny Cyclopaedia*, vol. xi, 87.

27. Hindley, "An Address," 499.

28. Wilson, *Imperial Gazetteer of England and Wales*, vol. i, 70.

29. Epstein, *Radical Expression*.

30. Anon., "A Reply," 213, 215.

31. Wallas, *Life of Francis Place*, 112.

32. On the British Association see Howarth, *The British Association for the Advancement of Science*. On the Manchester Literary and Philosophical Society see Kargon, *Science in Victorian Manchester*.

33. Sedgwick, *Life and Letters*, vol. i, 515.

34. *Ibid.*, 8.

35. Jefferson's *Notes on the State of Virginia* details the mineral, vegetable, and animal produced in Virginia, referencing a map of the state—exactly the standard practice before the stratigraphical work of Cuvier and William Smith (xx).

36. Cuvier, *Essay on the Theory of the Earth*, 24. Excellent sources for Cuvier and early nineteenth-century geology are Coleman, *Georges Cuvier: Zoologist*, and Rudwick, *The Meaning of Fossils*.

37. Dickens, *Bleak House*, 1.

38. Sedgwick, "Anniversary Address, 1830," 207.

39. Sedgwick, "Anniversary Address, 1831," 314.

40. *Ibid.*

41. Mill, "Literary Remains," *Collected Works*, vol. x, 144–45.

42. Sedgwick, *Discourse*, 12–13.

43. *Ibid.*, 19.

44. *Ibid.*, 104–05.

45. Galileo, "Letter to the Grand Duchess Christina," in *Discoveries and Opinions of Galileo*, 181.

46. *Ibid.*, 181.

47. Sedgwick, *Life and Letters*, vol. i, 379–81.

48. Conybeare, "Report," 388.

49. Phillips, *Illustrations*, part ii, v.

50. Phillips, *Illustrations*, part i, 12, 18–23.

51. Sedgwick and Murchison, "A Sketch of the Structure of the Eastern Alps," vol. iii, 351.

52. *Ibid.*, 313.

53. Sedgwick, "Anniversary Address, 1830," 205.

54. See Rudwick, *The Great Devonian Controversy*, and Rosenblatt, "Fossils and Myths."

55. Phillips, *Illustrations*, part ii, 36–37.

56. Sedgwick, *Discourse*, 31–32.

57. Ladd, *Victorian Morality of Art*, 53.

58. Ruskin, *Modern Painters*, vol. i, part I–II, preface to the 2nd ed., xxxiv, and vol. v, 387; see also Ball, *Science of Aspects*, 63.

59. Cannon, *Science in Culture*, 58–59.

60. Sedgwick papers at the Adam Sedgwick Geological Museum; compare, for example, notebooks 4 and 21.

61. See Rosenblatt, "Fossils and Myths," for an extended discussion of the language of systems geology, linking the work of Sedgwick and Murchison to that of E. Forbes and A. Gressly. We may add here a link to art and fiction via Alain de Botton's *How Proust Can Change Your Life*, where he writes: "What can replace a clichéd explanation of our functioning is not an image of perversity but a broader conception of what is normal" (99). He then suggests this rejection of stock characters and phrases by Proust may explain why he liked impressionist paintings, where an effort was made to render scenes with a distinct authenticity.

62. Sedgwick, *Discourse*, 8.

6

THE CHILD IS FATHER TO THE MAN

We come now "with a careful curiosity" to our second set of buckets drawn from scientific seas of 1832. As we do so, we may pause to pull together some leading finds.

We began with the Reform Bill. The agitation for reform set the pulse of the times. It is what was happening. It was the news above the fold on page one every day, with stories that often came amplified with the sounds of marching, charging feet. Reform was no minor adjustment in governmental niceties. It quickly opened to issues driven by the deepest forces of the day. With industrialization had come marked changes in where people lived, their relationships to their employers, and with one another. Old safety nets and guiding practices had been lost or rendered useless, if not counterproductive. New safety nets were called for, as in the Report of the Select Committee on Factory Children's Labour of 1832. New structures were called for, such as legal reform. And it all hinged on a more responsive Parliament.

We next examined classical scholarship. Greek and Latin had long been at the heart of schooling in England, with philology and rhetoric as handmaidens to biblical study and the proprieties of aristocratic rule. Scholars had long sought to integrate sacred and civil history, but with Mitford's *History* we can feel the emergence of something new, where civil history is claiming a greater share in that project. William Jones's Indo-European hypothesis illustrates the sweep of this historical vision, with Sanskrit as the oldest extant language and the mother of all subsequent tongues. Cuvier's geology would expand this project by adding the planking of natural history to a sacred and civil history of the world.

This grand project reflected a common set of goals and historical methods. It was a chronological project, fixing discrete items along a timeline; geological revolutions, catastrophic inundations caused by the raising of whole mountain chains, beat a steady stratigraphic pulse up to the modern era, the era of humankind. Then revolutions of the civil order take over: migrations, battles, inventions. History becomes a single story, a story beginning at a remotely distant point when the earth was first formed and carrying on through a succession of eons and eras, of floras and faunas, leading to the first humans, to Noah's flood and on to the rise and fall of languages, cultures, and civilizations.

Thirlwall and Grote deny this project. They deny the details could be trusted. More profoundly, they deny they were even of any genuine interest, either in civil history or, as Thirlwall and Schleiermacher lay out, in sacred history. The real story of antiquity is the story of the mind. For Thirlwall it was the imagination as it works to comprehend how the world had come to be the way it was. For Grote it was the subsequent march of critical thinking from poetic fancy to Herodotus and on to Plato and Aristotle.

Meanwhile, as we have just seen, Sedgwick's work is doing the same thing with natural history. He denies the link between natural and sacred history and more importantly denies the confidence geologists had placed in Cuvier's stratigraphical project, challenging the soundness of its data, such as discontinuities which dissolve as one moves horizontally into adjacent districts. As with Thirlwall, there was a grander tale to tell. Sedgwick's new systems geology sought the story of ancient communities, brought to life with topographical authenticity, with contours as settings that supported the plants and animals of a given epoch.

Along the way we discovered that the young Charles Darwin had studied with Sedgwick, a two-week tour in Wales in the summer of 1831, just after he had graduated from Cambridge. Later Darwin went out of his way to praise Sedgwick for all he had done. In an essay on Lyell's *Principles*, William Whewell explains the English school of geology in a way that helps us appreciate Darwin's debt. Whewell emphasized that in this school, "their disciples have been taught, their converts and teachers formed, with the hammer in their hands, with the knapsack at their backs, or at their saddlebow, in long and laborious journeys, amid privation and difficulty."[1] Sedgwick taught Darwin the feel of the hammer. With that tour Darwin matriculated into England's geological school. Some months later, when Darwin made his first forays into the wilds of South America as naturalist for the HMS *Beagle*, it was with geological hammer in hand.

As we turn to Lyell's *Principles* we should note that Henslow, the botany professor at Cambridge who was Darwin's mentor, gave him the newly published first volume as a parting gift for the voyage. And so, wrapped in the confidence that came from his introduction to the elite circle of English geology, Darwin settled into his duties as naturalist and geologist for the *Beagle* with a hammer in his hand and the *Principles* in his lap.

The *Principles* were an 1832 event for Darwin.

LYELL AND THE KALYDONIAN BOAR HUNT

In 1820 Charles Lyell entered Lincoln's Inn, the same year Thirlwall began. They seem not to have forged a notable friendship, however. We know, for example, that Thirlwall attended the annual meeting of the Geological Society in February of 1832, but Lyell's journal, which has a lengthy entry on that evening and speaks specifically of various literary men in attendance, does not mention Thirlwall.[2]

It is somewhat intriguing to speculate on why Thirlwall and Lyell did not become friends. We tread here on terrain similar to the Kalydonian boar hunt. Back in our discussion of Grote's *History*, we noted the example Grote offered of an early approach to myths where Ephorus, third century B.C., explained why Hercules had not been a member of this expedition; namely, that he was busy somewhere else.[3] Not a very promising precedent for us. Nevertheless, that these two bright and most able young scholars were in school together and yet did not connect is intriguing.

On the other hand, perhaps it is not so surprising. Thirlwall and Lyell were certainly cut from different cloth, as scholars and as individuals. Take the basic matter of place in society. Lyell's family was in what we refer to today as the "one percent." Charles had been born at the family estate in Kinnordy, a baronial mansion in the Highlands with acres and acres of land. The family also leased Bartley Lodge, a grand home (now a resort hotel) surrounded by eighty acres of forest and field near the New Forest in Hampshire. Lyell seems to have spent most of his childhood at Bartley Lodge, and split his married life between Kinnordy and a London residence. Meanwhile, Thirlwall's father was a cleric with a small living in Stepney in London's East End. Stepney grew rapidly across the nineteenth century and came to be known for its poverty, over-crowding, and political dissent.

Nevertheless, Thirlwall had long-lasting friendships with well-to-do schoolmates, such as Grote and Julius Hare. It was not just the money.

Perhaps it was more that Lyell was rather stiff and conservative. Take the Reform Bill. Lyell looked upon reform as a massive confiscation of property. In his journal he notes that for a man who has "several brace of borough appointments, each worth forty or fifty thousand pounds in the market, he may well cry out, and may easily persuade himself that the good desirable to the country is problematical." That this was meant as an understatement is made clear in another item from about this time. It turns out that Lyell composed an autobiographical sketch for his fiancé, whom he would marry in 1832. While recalling that he had changed schools when he was ten, he digressed to talk about Old Sarum, a pocket borough often cited in reform debates which, Lyell notes, Lord Caledon had recently purchased for the monies it would raise "little think-ing of the evil day which approached."[4]

Though Lyell stands alone on this matter for us, with Place, Thirlwall, Grote, Sedgwick, and Darwin all earnest supporters of reform, we can well imagine Thirlwall granting Lyell some space here; after all Sedgwick, Grote, and Darwin were each friends of Lyell. So maybe we are simply left with nothing more than matters of taste or happenstance, but while we are musing on such matters, there is another quality that seems relevant. Lyell was what Victorians called a "trimmer," a term borrowed from sailing, as in trimming one's sails as seas become rough. We might characterize trimming as exactly what Thirlwall did not do when he laid out his views on chapel attendance, or Place when listing the union's grievances, Grote when he pushed for a secret ballot, or Sedgwick when he used his presidential address to explain the need to see the Bible as a moral document and not a natural history of the earth.

There's an enticing passage from Lyell's autobiographical sketch that speaks to "trimming." Still a youngster, Lyell emulated his father and took up natural history, first plants and then insects. But he caught flak for these as less than "manly" pursuits, and he observes: "The disrepute in which my hobby was held had a considerable effect on my character, for I was very sensitive of the good opinion of others, and therefore followed it up almost by stealth; so that although I never confessed to myself that I was wrong, but always reasoned myself into a belief that the generality of people were too stupid to comprehend the interest of such pursuits, yet I got too much in the habit of avoiding being seen, as if I was ashamed of what I did."[5] This is quite a swirl of revelations: his early confidence in his own path and seeing those who would disapprove as stupid, and yet his dependence on the good opinion of others which leads him to be secretive, and finally shame at the need for others' good opinion.

Many years later the same mix of confidence, dependence on the opinion of others, and a resultant stealth in his views is evident on a key aspect of his work.

Lyell's *Principles* was bold and forthright on several matters, such as the age of the earth. It was central to his approach that we grant nature enormous quantities of time to accomplish the many events recorded in the rock and fossil records. But he was not so bold and forthright on everything.

Take the matter of the fixity of species. As we do so, we should remind ourselves of the proportions of this issue. Since Cuvier's master work early in the century, geologists had had a striking problem. The fossil record presented a vast succession of perfectly viable life forms that were now extinct, and while revolutions, catastrophes of extraordinary proportions, could explain how this might have happened, the question still remained: where did the new species come from? This was the mystery of mysteries. In an essay of 1852, Herbert Spencer estimated the total number of species of plants and animals which have lived over the course of life's history—some 10 million in all, thereby calling for 10 million distinct acts of creation. Unless we could determine a set of intermediate causes that could generate modified species out of existing stock, that is.[6] Fair enough. This gives us a feel for the proportions of the matter. Let's look at Lyell on this issue.

There is a striking letter from Lyell to his friend and critic William Whewell, written early in March of 1837, that takes us right to the heart of the matter. The letter opens by referring to an earlier conversation, one we may be sure was both earnest and energetic. It might have taken place a couple of weeks earlier at the annual meeting of the Geological Society where Lyell, as retiring president, would have turned the gavel over to Whewell, the rising president. But there is an urgency to the letter, so Lyell was probably responding to a more recent conversation; perhaps they had met at the Athenaeum or the chambers of the Royal Society. In any case, we are fortunate that Lyell returned to that discussion by post.

Lyell is chiefly concerned to rescue his work from the criticisms Sedgwick had leveled in his presidential address some *six years* earlier. Lyell was certainly sensitive to the good opinion of others. In the course of things he reminds Whewell of what Herschel had said to him in a letter of 1836, adding: "If I had stated as plainly as he has done the possibility of the introduction or origination of fresh species being a natural, in contradistinction to a miraculous process, I should have raised a host of prejudices against me, which are unfortunately opposed at every step to any philosopher who attempts to address the public on these mysterious subjects."[7]

We've jumped into another swirl here. We will come back to it all in a bit, including Herschel's letter; the crucial thing is Lyell's acknowledgment that he had kept his convictions hidden on a central issue because to do so would have raised a host of prejudices against him. The child is father to the man.

CHARLES LYELL—BY REFERENCE TO CAUSES NOW IN OPERATION

Geology is an historical science. Its data largely derive from the distant past, not from experiment. It might hardly seem significant to claim an historical science is, in fact, historical, yet it may be difficult to appreciate how history and science can be bound to one another. We live in an age where the image of science has been formed by the fantasy of space exploration, by the magic of medical technology, and the wonders of the cyber world. There is little hint here of the historian's visage. If history cannot be reproduced in a flask, then how can it be analyzed scientifically? Perhaps there is not a genuinely historical face to the historical sciences, sciences like geology and evolutionary biology.

Such a tension between science and history may be traced at least as far back as Plato. There is a passage early in the *Timaeus* which forbids the use of "this." True knowledge requires a move from the particularity of an item, its "this-ness," toward how it participates in universals. Insofar as history seeks to explain the unique, the particularity of the past, it is not true knowledge of the sort provided by science.[8]

Two millennia later, R. G. Collingwood distinguished between history proper, which is about human thoughts and actions, and pseudo-history, which merely arranges events and relics along a time line. Pseudo-history is an external kind of knowing, characteristic of the natural sciences.[9]

As much as I respect Collingwood, this perspective will not do. We have already seen rich parallels between Sedgwick's revolutionary work and the brilliant classical history of his friend Connop Thirlwall. For both Sedgwick and Thirlwall there was a shift from a universal, chronological history to a regional history. For both there was a shift from the telling particular to a broader unfolding of changes in response to changing conditions, be they mythopoetic constructs or fossil forms. For both there was a shift from a vertical arranging of events and relics along a time line to a richer horizontal gaze where the historian evoked an ancient landscape or "mindscape" that shaped life's expressions, be it fossil or myth. These challenges to the prevailing approach to history were not the result of new data. They were a rethinking of what sound history was about. Sedgwick's evocation of ancient landscapes and related "ecosystems" was, like Thirlwall's trace of the evolving sensibilities of the ancient storyteller and his audience, a new take on the deeper story to tell. The substantive core of Sedgwick's work, as with Thirlwall's work, was genuinely a matter of historiography.

Now let us turn to the work of Charles Lyell. Lyell's *Principles* was extraordinarily successful. It quickly secured Lyell a distinguished place in the scientific community, a stature time has amplified. There is a considerable body of commentary on Lyell, far more than any other figure in the history of geology. Most discussions of Lyell's work pay little attention to the elements of historiography, but let us see where an eye on historical criticism will take us.

SKEPTICISM AND SOUND
HISTORICAL INVESTIGATION

The English school of geology, so richly evoked by Whewell, "with hammer in hand," had forged its character around the formative work of William Smith and Georges Cuvier early in the century. The unity of this school was Lyell's problem. His *Principles* was not a drawing together of the many local studies of his colleagues. It was a drawing apart. There was a difference between their geology and his, a difference in principle.

The first volume of the *Principles* quickly settles into a "Historical Sketch of the Progress of Geology," starting with the earliest speculations of myth and legend. Like Grote, Lyell was not interested in their content, but in what constitutes progress; that is, the character of effective analysis. He did not reduce the ancients' tales to credible proportions, nor did he discredit them with fieldwork. He simply dismissed them as superstitions of the yet uncultivated. They were of an age when history had yet to sever itself from poetry and myth.[10]

The parallel is striking.

Early on, people grappled with how to make sense of phenomena that were out of the ordinary, such as eclipses, earthquakes, and great floods, linking them to the gods. In time, such dramatic events and catastrophes would be reconciled to the same forces that rule everyday life. For Lyell, as for Grote, in history and in science, there has been a progress of rational criticism over credulity and of present experience over superstition.

However, even as rational criticism grew in other domains, something acted to retard right reasoning in geology. Something disposed geologists to disregard the known course of things and allow the present to be but half-brother to the past. Lyell proposes that it was the mistaken idea that the earth was not very old. By compressing vast reaches of time into mere thousands of years, such histories of the earth gained "the air of a romance."[11] He went on to explain himself by borrowing from the legend of the seven sleepers from Gibbon's *The Decline and Fall of the Holy Roman Empire*, which tells of the shock of the seven after they had slept in "Rip van Winkle-fashion" for generations. This simple tale nicely illustrates how the sudden juxtaposition of endpoints might give the appearance of dramatic change when in reality it was the ordinary course of things played out over long stretches of time.

Such is the heart of Lyell's take on things: the facts, as they present themselves, cannot be trusted. They give the appearance of dramatic change and discontinuity when there may well have been nothing but the ordinary course of nature. The evidence is not a rational deposit of the past. It is not a carefully kept archive of the succession of events, the raising of mountains, the coursings of rivers, the ebb and flow of a vast diversity of plants and animals. It is only a random sampling of relics and scar tissue, itself subject to the wear and tear of physical and organic forces.

How should we understand this core postulate in Lyell's work? It is certainly not a simple statement of fact. One could not go out and directly observe whether or not it was so. Nor is it an analytical truth, drawn from clear and distinct truths and warranted by reason itself. It is a maxim, a matter of insight, and it serves as a guide to how we may or may not interpret what is before us. It is a principle of historical criticism of the sort we have traced in Grote's *History*. Both denied the legitimacy of the text.

Skepticism is a coat of many colors. In his "Autobiography," Darwin tells a delightful tale of Sedgwick's skepticism. It is the summer of 1831 and the good professor has stopped at Darwin's home in Shrewsbury on his way to north Wales, having agreed to take the young Darwin along for some fieldwork. That evening, Darwin tells Sedgwick of visiting a nearby gravel pit and of a fossil shell of a distinctly tropical sort which a worker claimed had been found there. Sedgwick doubted it at once. We can readily imagine the embarrassment Darwin felt at the ease with which Sedgwick dismissed his find. Darwin acknowledges in the autobiography that he checked it out, and the workers confessed they'd played a practical joke and the fossil was not from the pit.[12]

This, however, was not the sort of skepticism which colors Lyell's *Principles* or Grote's *History*. Theirs was more thoroughgoing and formative.

Then there is the skepticism of Montaigne. This, too, is a tint different from Lyell and Grote. Montaigne scoffed at idealism. His was the prudence of the practical man of affairs with a keen eye who distrusted the follies of reason.[13] Such down-to-earth, jaundiced-eye, back-room stuff will not do for Grote. He was an idealist. And while he was not a religious man, his philosophy was, for him, a religion in that important sense of Carlyle's when he said of religion that it was not a matter of creed professed, but "the thing a man does practically lay to heart, and know for certain, concerning his vital relations to this mysterious Universe, and his duty and destiny there."[14]

Grote's skepticism was recognized for its boldness. For John Stuart Mill his *History* was remarkable for its "unsparing manner," while to the historian, Henry Milman, Grote's "remorseless" skepticism had led him to discard too much. In replying to Milman, Grote explained that he had not denied that there may be matters of historical fact within the legends, only that we may know what they are.[15] It was an "epistemological" skepticism.

Let's see what hue of skepticism colors Lyell's work.

THAT LETTER

We may now take a closer look at that letter Lyell wrote to Whewell. You will recall that Lyell was explaining how Sedgwick's criticisms had been unfair. For example, Sedgwick distinguished the immutable primary laws of nature from the eminently mutable results of their combination. To assume, as Lyell would have us do, that volcanoes have acted throughout the indefinite reaches of the earth's history with a uniform intensity was a "merely gratuitous hypothesis, unfounded on any of the great analogies of nature, and I believe also unsupported by the direct evidence of fact."[16]

Lyell responds to this charge in his letter to Whewell: "I expressly contrasted my system with that of 'recurring cycles of similar events.'... I never drew a parallel between a geological and an astronomical series or cycle of occurrences."[17] What is going on here? Why does Lyell jump to cycles and a connection between geology and astronomy? It turns out we need to step back just a bit. Sedgwick's predecessor as president of the Geological Society was William Fitton, whose 1829 presidential address had sung the praises of the doctrine of Hutton and Playfair, as one where "the forces from whence the present appearances have resulted, are in Geology, as in Astronomy and general Physics, permanently connected with the constitution and structure of the Globe." Furthermore, though the operations of nature are of great complexity and variety, when viewed across vast eons acquire a "sort of uniformity."[18]

There is a grandeur to this view, as vicissitude gives way to uniformity, the Laplacean uniformity of the heavens, but it is not Lyell's view. Sedgwick has held Lyell guilty of Huttonian sins. Hutton, Playfair, and Fitton rested geological uniformity on the character of nature's dynamics. For them uniformity was an empirical claim, based on the structure of things. But for Lyell things were different.

Lyell asks Whewell to consider a related charge Sedgwick had leveled that he, Lyell, had ruled out a priori that there had been any great catastrophes in the past. In Sedgwick's fine words: "And what is this but to limit the riches of the kingdom of nature by the poverty of our knowledge; and to surrender ourselves to a mischievous, but not uncommon philosophical scepticism, which makes us doubt the truth of what we do not perfectly comprehend."[19] Instead, Lyell explained: "I argued that other geologists have usually proceeded on an arbitrary hypothesis of paroxysms and the intensity of geological forces, without feeling that by this assumption they pledged themselves to the opinion that *ordinary forces and time* could never explain geological phenomena."[20] In short, before you claim things were different in the past, you should first show you couldn't explain the evidence by relying on ordinary forces.

Then you get the punchline: "The reiteration of minor convulsions and changes, is, I contend, a *vera causa*, a force and mode of operation which we know to be true." The phrase "*vera causa*" is a double whammy; it invokes both Isaac Newton, who used the expression in his first rule of reasoning in his masterwork, the *Principia* and John Herschel, who used it in his *Preliminary Discourse*. Both uses seek to distinguish real forces from mere hypotheses, precisely the distinction Lyell is hammering on here.[21] We can measure the intensity of the ordinary forces of nature: the coursings of rivers, the uplift caused by earthquakes, the rate at which species become extinct. These are *vera causae*, and they should be the basis for geological explanations. That the intensity of forces has been different at times in the past, Lyell adds, "may be true; I never denied its possibility, but it is conjectural."[22] An echo of Grote's reply to Milman, Lyell's skepticism carries an epistemological tint.[23]

The point was not what actually happened. We have no access to that. The past, especially the distant past, is necessarily something we invent. The challenge is to constrain our inventiveness, so that we are genuinely reasonable—a challenge both Grote and Lyell met with "present facts."

Going back to the letter to Whewell, Lyell moved from Sedgwick's criticisms to that most controversial subject: the fixity of species and the fossil record. The year before Herschel, who was in Cape Town, South Africa, busily mapping the heavens of the Southern Hemisphere, had written Lyell, praising his approach to geology and drawing particular attention "to that mystery of mysteries, the replacement of extinct species by others."[24] As we noted, once we accept that the fossil record is showing us anatomically viable species, then we need to explain why they are not around anymore. This is the problem of extinction, which Cuvier solved by invoking paroxysms of great violence, but once you accept that species become extinct and are replaced by new forms, the issue of where they came from is front and center. Cuvier offered no hypotheses here. The mystery of mysteries was the price he paid for his brilliant analysis.

Herschel then went on to offer his own view that "in this, as in all his (the Creator's) other works, we are led, by all analogy, to suppose that he operates through a series of intermediate causes, and that in consequence the origination of fresh species, could it ever come under our cognizance, would be found to be a natural in contradistinction to a miraculous process—although we perceive no indications of any process actually in progress which is likely to issue in such a result."[25] Herschel is convinced that there is some natural explanation for species change and though Lyell agrees, he hid his views from the public, not just in the 1830s but for decades to come. Even after Darwin had published *On the Origin of Species* in 1859, Lyell waffled about whether humans had evolved in his *Antiquity of Man* of 1863. Darwin was so upset he wrote to Lyell: "I will first get out what I hate saying, viz that I have been greatly disappointed that you have not given judgement & spoken fairly out what you think about the derivation of Species . . . I had always thought that your judgement would have been an epoch in the subject. All that is over with me."[26]

SECRET HANDSHAKES

Later in the century, in an essay on Darwin's *On the Origin of Species*, T. H. Huxley linked the work of Darwin, Lyell, and Grote. Lyell's approach to geology had paved the way for Darwin's understanding of the history of life. Lyell, Huxley went on to explain, had brought home to any reader the great principle "that the past must be explained by the present," and further he went on to assert that this was true of all sound history: "Grote's *History of Greece* is a product of the same intellectual movement as Lyell's *Principles*."[27]

We agree that present facts are crucial to these three historians. But having established that Grote and Lyell are virtually congruent triangles, are we now charged with the burden of showing that Lyell came to his understanding by way of Bentham or his circle? We know Lyell was a friend of the Grotes and that they would go off on little geological hikes together. But that Lyell was a friend does not mean that Grote had initiated him into the Benthamian sect, complete with secret handshake. Lyell had come to London early in the 1820s and set up shop as a lawyer. He was likely to have been familiar with Bentham's views on the law, but we have also seen that he was rather conservative and not a likely acolyte of philosophical radicalism. It is altogether plausible to see Grote as having picked up and honed his approach to analysis within the Benthamian circle, just as it is reasonable to see the influences our Trinity scholars would have had upon one another, but such connections should not set the expectation that influence is a direct handoff. Lyell could have come by his thoroughgoing skepticism by many another route.

One such route might have been writings from the Scottish Enlightenment, which was marked by a number of accomplished historians, including David Hume, Adam

Smith, and Adam Ferguson. Let's take a look at Ferguson's *Essay on the History of Civil Society*, which begins with a criticism of the rationalist philosophers, such as Locke, who have speculated on the origins of society "in some imaginary state of nature."[28] Instead of such hypotheses, Ferguson turned to empirical qualities, concepts richer in detail, if more vaguely defined, closer to practice, and above all tangible, intellectual realities we can almost taste. He turned to habits, passions, sentiments, and dispositions. These were the proper analytical units for a study of the human condition.

Ferguson saw himself as a naturalist who, when he examines any species, "supposes that their present dispositions and instincts are the same which they originally had." In applying this notion of an essential uniformity over time to humanity, Ferguson examined the range of the human condition across the globe in the late eighteenth century—from the rude society of the American Indian to the polite society of the Parisian salon. This lateral progress displayed in the plane of present experience, was then drawn into a line and thrust into the depths of history.[29]

This same sort of analysis was used by Thomas Jefferson in a succinct passage from a letter written in 1824. Jefferson invites the reader to begin with the "savages" of the Rocky Mountains, where you would find them "in the earliest stage of association, living under no law but that of nature." But moving eastward, you "would next find those on our frontiers in the pastoral state, raising domestic animals." Then there are "our own semi-barbarous citizens, the pioneers of the advance of civilization," and then, finally, "the most improved state in our seaport towns." This is then brought to a close with this observation: "This, in fact, is equivalent to a survey, in time, of the progress of man from the infancy of creation to the present day."[30] This is the naturalist's past. It is the past residing in the present.

PRESENT FACTS

In the face of evidence that was riddled with uncertainty, the only rational recourse was the present order of things. The analogy between past and present would serve to curb speculation and keep things close to what was actually possible. The challenge for Lyell, reflected in the subtitle of the *Principles*: "An Attempt to Explain the Former Changes of the Earth's Surface, by Reference to Causes Now in Operation," was to capture the complexity and subtlety of the present course of nature.

The supposition of uniformity that Ferguson and Jefferson had made was of a rather Huttonian, structural sort. The present is framing what the past had been like. But Lyell's take is more subtle, as the streets of Rye and Aldeburgh show us.

In the South of England, not far from the site of the ancient battle of Hastings, is the lovely town of Rye. Rye has a history reaching back into the dim recesses of the medieval past. It was, for example, one of the "cinque ports"—not exactly, but essentially so, for it was part of that scheme to provide the British monarchy with a navy to defend its shores and to wage war. There were seven cinque ports, five smaller ports and two cities, Rye and Hastings. These seven towns were charged with providing a certain number of ships and men in case of war. In return they received various privileges, most notably that of being a safe haven for escaped serfs, thereby providing them with a labor force.

Rye's participation in England's military affairs would reach its zenith in 1588 with England's defeat of the Spanish Armada. On the eve of that great battle, the small British fleet hid, unseen, in Rye's harbor.

There is lots to see and learn about in Rye, such as the smugglers who used the Mermaid Inn, but the fact of Rye's maritime history is especially intriguing because there is no water to be seen from Rye today; the town is landlocked. Where there used to be a busy harbor on the English Channel, there is now a pasture with grazing sheep. You can even walk out of the city, down a steep slope, and walk across ancient ocean floor to Winchelsea, another de-harbored port city. Or you can set out across the fields to Camber Castle, which used to be an island fortress guarding the entrance to the harbor. Just amazing!

Figure 6.1 depicts an undated map perhaps from the early sixteenth century, where Rye is on a small island to the left. Note that the map indicates tracts of land which had been "inned," or claimed from the sea in the twelfth through fifteenth centuries.

And Figure 6.2 shows another map, this one from 1595, where the changes are clear. The water has continued to recede. By the 1970s the channel was about two miles away from Rye, and almost all of the marsh had been inned.

But what happened? How could Rye lose its harbor?

One possibility is that the ocean has fallen the fifty or hundred feet it would have taken to drain Rye's harbor. If this were so, however, we would expect to find other harbors had been affected in the same way; this is not always the case. In fact, less than one hundred miles from Rye is another ancient port town, Aldeburgh: lovely and quaint, with clapboard homes along the North Sea coast and a main street with old shops. Walking along the main street down toward the sea, the last cross road, the beach road, is 7th Street. But you have not left 5th and 6th streets. The streets behind you are 9th and 8th streets. Assuming no one would lay out a town beginning with 7th Street, what happened to streets one through six? The answer is not blowing in the wind; it is lying in the sea, for there, off the shore, is the top of the old city hall. The sea has steadily crept up on this old town, and half of it lies buried by sea and sand. Aldeburgh has long struggled with an encroaching sea. So much so that though in Tudor times its bustling port and ship-building led it to be granted two MPs, by 1832 the population had so dwindled that they were taken away with the Reform Bill.

So if the sea is coming in at Aldeburgh and going out at Rye, then it's not the sea level that changed so markedly. Lyell discusses Rye and Aldeburgh, along with other examples, to illustrate the flux of uplift and subsidence, which, along with changes in sea level, are constantly altering coastlines.[31]

There was more to this than initially meets the eye. There's an interesting puzzle at this time concerning "erratics," large boulders distinctly out of place that nicely illustrate the edge in Lyell's work. In 1831 the Rev. William Conybeare, whom we have met once or twice already, wrote a lengthy article teasing out items that sharply defined the differences between his approach to the earth's history and that recently laid out by Lyell in his *Principles*.

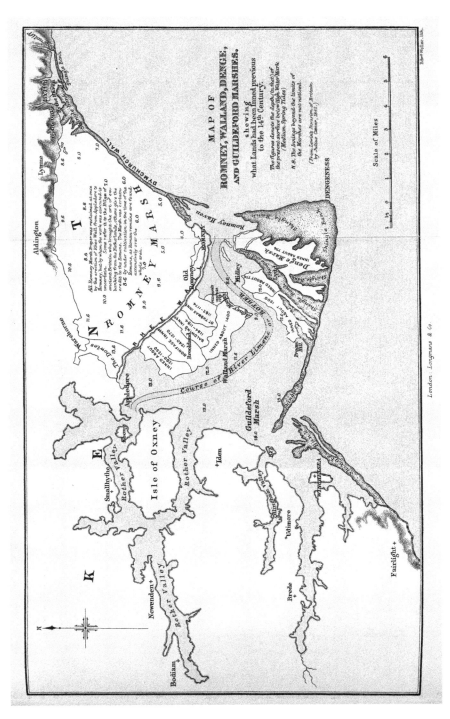

FIGURE 6.1 Map of Rye, Romney Marsh, from *Cinque Ports* by Montagu Burrows, early sixteenth century.

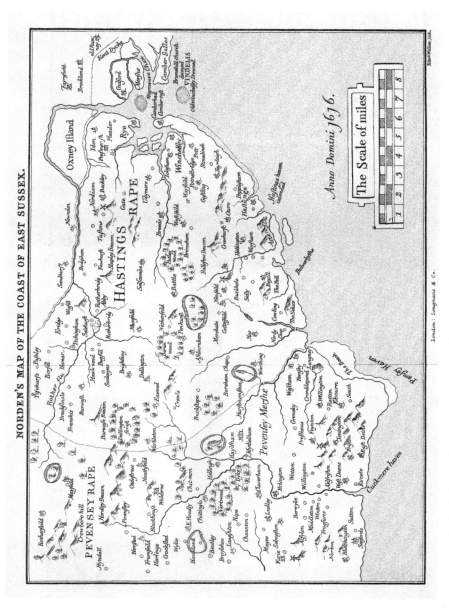

FIGURE 6.2 John Norden's map of 1595 of the coast of East Sussex, from *Cinque Ports* by Montagu Burrows.

In the course of things Conybeare pointed to the erratics scattered across the broad flat plain of the Luneberg Heath in the north of Germany. These were "vast blocks of granite, often as large as small cottages" completely unlike the bedrock gravels, chalk, and flint of the heath and evidently derived from mountains in Norway. But how did they get there? It was one thing to allow weathering and gravity to break off large boulders and carry them downhill one way or another to the floor of the Baltic Sea, but what feature of Lyell's "causes now in operation" could transport them uphill from seafloor onto the shores of Germany and still further carry them inland against the current of any river, which all flow north down to the sea?[32]

Conybeare was convinced that only something out of the ordinary course of nature, like a substantial flood, could explain them, but Lyell had an interesting hypothesis up his sleeve. Start with glaciers, rivers of ice that work their way from mountain top and hillside down to the sea. As they hit the standing water of a bay, an ocean, or in this case the Baltic Sea, they move out over the water until the weight of the ice is greater than the bonds that hold it together and so an iceberg is "calved." Such icebergs, hundreds of which may be seen drifting about, especially in arctic waters, carry debris which had been gouged out by the "flow" of the glacier down to the sea.

These icebergs would drift until they melted or beached against a shoreline. This alone would not explain the erratics, because the debris from an iceberg may be carried close to the shore, but is never deposited upon dry land. Coastlines, however, are constantly shifting, as in Rye and Aldeburgh. Erratics are simply markers of former coastlines. They flag where icebergs had beached. At some point the shoreline had encroached, Aldeburgh-like, upon the present shore. The iceberg beached and melted. Then Rye takes over, and the coastline moves out. Where Conybeare had seen the clear mark of some catastrophe, a flood great enough to sweep huge chunks of granite across the Baltic, Lyell saw nothing but the play of ordinary forces: a bit of subsidence here; a drifting iceberg beaching, melting, and dropping its burdensome pack; and finally a bit of elevation—all causes now in operation without hint of massive disruption.

We should add that clever as it was, Lyell's iceberg drift was not the last word. Before the close of the decade a new hypothesis was put forward by Louis Agassiz and others— an Ice Age. Soon both Lyell and Conybeare saw erratics as leading evidence for major climate change at different times in the past, though Agassiz envisioned the onset of the Siberian-like freeze as catastrophic and Lyell urged a more gradual process.

SO WHAT ARE THE FACTS?

It was central to Lyell's argument that such changes evident in present facts be taken as the scale of change for reasonable conjectures about the geological past. There were no great cataclysms at Rye or Aldeburgh. People continued to go on about their business. This leads directly to the whole matter surrounding discontinuity. Here's how Sedgwick put it: "If the principles vindicated in Mr. Lyell's work be true, then there can be no great violations of continuity in either the structure or position of successive formations."[33] Yet clearly there were such discontinuities in the rock record and the fossil record. Did Lyell refuse to face the facts? Yes and no. He accepted the appearance of discontinuity, but denied that such appearances implied the violence others had taken them to imply. Like Rye and Aldeburgh, change could happen so slowly you would hardly notice it from day to day, month to month, or year to year.

In a chapter on the action of rivers, Lyell used present facts to show that discontinuity is not a simple matter of fact. He describes the flooding of the Mississippi, where "rocks and trees are hurried annually, by a thousand torrents . . . together with the wreck of countless forests and the bones of animals which perish in the inundations." And then he drives home the moral of the tale: many geologists, when they see such debris heaped confusedly "read in such phenomena the proof of chaotic disorder, and reiterated catastrophes, instead of indications of a surface as habitable as the most delicious and fertile districts now tenanted by man."[34]

Though present facts are immediate, the context they provide is curiously abstract. Instead of a regional study, like all of the significant papers of this era—be it Sedgwick on Wales, Phillips on Yorkshire, or Sedgwick and Murchison on Devonshire—Lyell pulls isolated bits from across the globe. In a short section on springs, for example, Lyell sampled more than thirty different sites, ranging from Trinidad to Iceland to the Dead Sea, teasing out from them a scale of intensities similar to Ferguson's scale of civil societies. In one reflective passage Lyell calls this account of the operations of nature "the alphabet and grammar of geology."[35] We can thus appreciate the irony that Lyell, who is often hailed as the figure who put geology onto its empirical feet, was charged by Sedgwick with refusing to face the facts.

The fact is, Lyell's empiricism rested on a denial of the facts—as they had been taken. The geological record was as obscurely connected to the actual past for Lyell, as the connection between ancient legend and historical events for Grote. The key to sound geology was finding some way to dispel deception and establish a connection between the evidence and pointable reality.

Just as Thirlwall and Grote were responding to Mitford and not one another, so Lyell's analysis took off from Cuvier, not Sedgwick. One illustrative example is the work that occupied Lyell across 1832, his study of Tertiary strata. Volume two of the *Principles* had appeared in December of 1831. Shortly thereafter he began a series of lectures at the Royal Institution, Faraday's home, and a fuller set of lectures at the newly opened King's College. A busy year, but the fullest charge on his time was his study of how best to structure the Tertiary.

As these strata came from the more recent past, they would "have been less mutilated by the hands of time."[36] Yet even here one needed to be careful. The fossil evidence could not be used as it had been. Cuvier had focused on tertiary strata in his study of the Paris basin, and used the remains of large land animals to trace a number of great floods. That such large animals would become fossilized, a process involving their remains coming to rest in the shores of lakes and seas, suggested catastrophes where the land was swept by vast inundations. How else would giant megatheria and their like be fossilized? Further, the advance of life from the simplicity of its earlier forms to the complexity of existing animals was more clearly delineated with land animals than creatures of the sea, like mollusks and their ilk. The number of quadrupeds is far more limited and their characteristics far more distinctive, and so the march of progress would be clearly delineated.

Lyell stood Cuvier's judgment on its head. He denied the soundness of inferences from land animals and rested his work on the marine life Cuvier had set aside. Fossils are the remains of individuals which had died and come to rest where conditions were favorable for preservation. As these are chiefly accumulating sea beds, for all but marine life to become fossilized was essentially a matter of chance. The remains of a megatherium did not imply a great flood. It only implied the play of fortune. Lyell opted for the sturdier frame of such creatures as mollusks, whose lives were played out in just those accumulating sea beds. The gradual changes in their large, amorphous populations were more likely to reflect the true tempo of geological change.[37]

There is a curious quality to this argument. The value of the myriads of shelled sea life for Lyell was that their habits did not get in the way. And what did they tell him? Did he ferret out the contours of ancient seascapes and the ecosystems they supported, like Sedgwick and Phillips? No. He divides the fossils into two classes: those that were essentially similar to living species, and those that were not. He then sifts through the fossils of entire deposits and determines the percentage of "living" forms. He simply sought a relative measure of the degree to which deposits approached the modern epoch. Comparing three thousand fossils found in tertiary strata to five thousand living species, Lyell finds the newer Pliocene has some 90 to 95 percent living fossils, while the older Pliocene has only 35 to 50 percent, the Miocene about 20 percent, and the Eocene 3 percent. In all this the character of any particular deposit has been lost in the shuffle. Indeed, he acknowledges that two deposits, say one from India or South America and the other from the Paris Basin, might have no particular species in common, but belong in the same group by virtue of being equally removed from the present via this look-alike contest.[38] Lyell's distrust of the fossil evidence led him to a far different point from topographical geology, where the criticism was that Cuvier had not fully enough evoked ancient landscapes and their flora and fauna.

As we come to the close of our study of Lyell, three sets of conclusions have been earned. There is first the general parameters of his geology. The *Principles* deeply influenced the course of geology and the historical sciences more broadly. It constituted a new and powerful program for research in processes of change. Also, he did more than emphasize that the earth was vastly older than a biblical ten thousand years; he laid out a program for how to tap into that vast reserve of time to make mountains of molehills and erode, in turn, whole mountain ranges one grain at a time. He offered more than a vista. He ventured into these tracts of time and staked claims for the significance of gradually shifting shorelines, tumbling streams, and global climates that varied at an almost indiscernible rate. And even his foray into a more historical geology, dividing the Tertiary into four phases delineated by their biological distance from the present, still stands largely intact.

A second set of conclusions concerns the rich parallels between Grote's *History* and Lyell's *Principles*. Three broad aspects resonate deeply between these influential works.

- Both were topical; that is, instead of laying out a chronological account, as had other classical historians and geologists, they presented catalogs, be it different sorts of myths or different agencies of change: igneous, aqueous, organic.

- Both shifted historical analysis from reconstructing a sequence of events to processes of change and they sought the measure of these processes in "present facts."
- Both anchored their approaches to history in an epistemological skepticism that focused on what it was we could reasonably know of the past by taking measure of the present. Whatever the past might have been like, the most rational approach was to assume a basic continuity over time in the processes of change.

Finally, there is the conformity in the work of Thirlwall and Sedgwick, Grote and Lyell, and of the four of them together in their critique of the Mitford–Cuvier approach to history. The brilliance of these four historians as they teased out the meaning of fossil and of myth stands alone, but the parallels point to deeper currents in early Victorian thought and sensibilities.

Here is a hint of these currents from the pen of Samuel Taylor Coleridge, where he contrasts the Bible and that shallow, abstract sort of thinking he variously called materialist, empiricist, atomist, and Scottish. It was for him the mechanic philosophy of the dead: "And in nothing is Scriptural history more strongly contrasted with the histories of highest note in the present age, than in its freedom from abstractions." And he continues: "How should it be otherwise? The present and preceding century partake in the general contagion of its mechanic philosophy."[39] So did Grote's *History of Greece*; and so did Lyell's *Principles of Geology*.

NOTES

1. Whewell, "Lyell—Principles," 180.
2. Lyell, *Life, Letters*, vol. i, 112, 372–73.
3. Grote, *History*, vol. i, 142.
4. Lyell, *Life, Letters*, vol. i, 9, 322–24.
5. *Ibid.*, 17. See also Allen, "Sir Charles Lyell," 591–609.
6. Spencer, "The Development Hypothesis," in *Essays*, vol. i, 1–7.
7. Lyell, *Life, Letters*, vol. ii, 5. See also Rudwick, "Introduction," to Lyell, *Principles*, xxx. We find the same tone in a note from Lyell to Darwin twenty years later. Appreciating how thoroughly Darwin has thought things through, Lyell shares his misgivings that man, too, has evolved, along with frogs and dogs, bushes, birds, and beasts: "It is this which has made me so long hesitate always feeling that the case of Man & his Races & of other animals & that of plants is one & the same & that if a 'vera causa' be admitted for one instead of a purely unknown & imaginary one such as the word 'Creation' all the consequences must follow" ("Letter of Oct. 3rd, 1859," in *Correspondence of Charles Darwin*, vol. vii, 1858–1859, 340).
8. Plato, *Timaeus and Critias*, 68; see also Gould, *The Panda's Thumb*, 27–28, and Hooykaas, *Natural Law and Divine Miracle*, 143 and passim, where he writes that geology is scientific, not historical, and that history is restricted to humankind.
9. Collingwood, *Idea of History*, 302.
10. Lyell, *Principles*, vol. i, 8.
11. *Ibid.*, 78–79.

12. Darwin, *Life and Letters*, vol. i, 25.
13. Emerson, "Montaigne; or the Skeptic," in *Representative Men,* 149–84. Here is a lovely line of Emerson's: "The astonishment of life, is, the absence of any appearance of reconciliation between the theory and practice of life" (177).
14. Carlyle, *On Heroes*, 3.
15. Mill, "Grote's *History of Greece*," 348; Milman, "Grote's *History of Greece*," 119; and Grote, *History*, vol. i, 394 (see note at bottom of the page).
16. Sedgwick, "Anniversary Address, 1831," 301.
17. Lyell, *Life, Letters*, vol. ii, 3.
18. Fitton, "Anniversary Address, 1829," 133–34. Fitton quotes a passage from John Playfair's *Illustrations of the Huttonian Theory of the Earth*, comparing Hutton's geology and the motions of the heavenly bodies, in both there is continual change bounded such that deviations in one direction "must become equal to deviations from it on the other" (440).
19. Sedgwick, "Anniversary Address, 1831," 301.
20. Lyell, *Life, Letters*, vol. ii, 3.
21. Newton, *Mathematical Principles*, vol. ii, 398, and Herschel, *Preliminary Discourse*, 197. In an earlier letter to Whewell, December 29, 1831, Lyell notes a curious hypothesis about an epoch when the earth was covered with a peculiar set of minerals, such as marl, salt, and gypsum, an epoch completely unlike the existing order of things. What he finds still more curious is this: "I am not aware that any science was ever before in such a state that the inexplicable mystery of a phenomenon was a subject of triumph to a large proportion of its cultivators . . . In other sciences I fancy that explanations by known causes has always been a victory rather than a defeat" (ms.116 of Whewell papers). Lyell uses *vera causa* in just this sense of a known cause.
22. Lyell, *Life, Letters*, vol. ii, 3.
23. Lyell *echoes* Grote because we have examined classical studies before geology. It is an echo for us. Though Grote had completed, or virtually so, his study of Greek mythology by the time he was elected to Parliament in 1832, the first volume of his *History* was not published until 1844, well after Lyell's letter to Whewell.
24. Herschel, "Letter to Lyell, Feb. 20, 1836," 226–27. Not only was this letter published in Babbage's work, but that same month, July of 1837, it was the subject of a meeting of the Geological Society. See Cannon, "The Whewell-Darwin Controversy," 378.
25. Herschel, "Letter to Lyell," 226–27.
26. Desmond and Moore, *Darwin's Sacred Cause*, 329–30, where they cite the *Correspondence of Charles Darwin*, vol. xi, 403. They also cite a letter from Darwin to the Rev. Charles Kingsley, where he had written of his expectations that Lyell would be taking on the relation between men and other animals in his new book (*ibid.*, 318).
27. Huxley, "On the Reception of the Origin of Species," in Darwin, *Life and Letters*, vol. i, 543.
28. Ferguson, *Essay on the History of Civil Society*, 3.
29. *Ibid.*, 3–4.
30. Jefferson, Correspondence, "Letter to William Ludlow, Sept. 6, 1824."
31. Lyell, *Principles*, vol. i, 274–78.
32. Conybeare, "An Examination," 196.
33. Sedgwick, "Anniversary Address, 1831," 306.
34. Lyell, *Principles*, vol. i, 189–90.

35. *Ibid.*, vol. ii, 190.
36. *Ibid.*, vol. iii, 62.
37. *Ibid.*, vol. iii, 31–34, 44–49.
38. *Ibid.*, vol. iii, 58.
39. Coleridge, *Statesmen's Manual*, 321.

7

IS THE MAP ANY GOOD?

It was a large, spare, rather bleak hall with folding chairs scattered in small clusters and dozens more leaning, folded against the walls. Ernst Gellner was the lecturer and this was the first of a series about leading sociological theories.

In time the room began to fill with students chatting away about their summer holidays and plans for the coming year. By the time the lecture was supposed to begin the room was full and quite noisy, but no sign of Dr. Gellner. A further ten, fifteen minutes went by. Students were still chatting. At last he showed up, making his way to the platform at the head of the hall. There was a quick shuffling of chairs and an expectant silence filled the room.

Gellner walked to the wall at the back of the platform, set his briefcase and cane down, took off his raincoat and folded it, and set it on the floor. He walked to the lectern, rested his cane against its side. Still no word, not even a look across the room to catch the eyes of any of the assembled. He then cradled his head in his hands, elbows resting on the lectern for what seemed a long time.

He began to talk. No idle chat. No welcome to the start of a new year. No apology for being late. Just the story.

His first fieldwork had been with a nomadic group living on the northern flanks of the Sahara in Morocco. It was very intense. He had been living with them and carrying his weight as an adult within the community, while at the same time absorbing cultural practices, noting occasions where behavior was surprising, attending to all sorts of nuances and trying to make sense of it all. After two or three weeks, he would need a break.

On one occasion he caught a ride to mountains. He hiked for several hours and all was well; it had been a lovely day. Then, late in the afternoon, it was time to start making his way back toward the highway. He took out his map, scanned the horizon, and decided that a given peak over "there" must be this one "here" on the map, and that other one must be this one "here." This, in turn, meant that he was in this place and he had a relatively short hike to get to the highway.

It didn't work.

So he took out his map again and re-examined the landscape and how it might correspond to the map. He decided he was really "here" on the map and so needed to go in this other direction for the highway.

Again it didn't work.

He had only brought a small day pack with a little food and water and no extra clothing. Things were serious. If he did not find his way out of the mountains he might well die of exposure. Temperatures plunge quite markedly at night in arid climates. He unfolded the map once again and carefully examined the terrain and the contours of the map, looking for the right match. He tried another route, but it too failed.

At this point he made a critical decision: the map was no damn good. For whatever reason, it was no good as a guide. He was on his own. He looked about and chose the direction he thought most promising. He lived to tell this tale, so we know he'd made the right choice. But that wasn't the moral of the tale. The moral was that the world is a very different place when you come to see that the maps are no good.

With a map, rational practice is essentially a matter of conjecture. The guidelines for things that are important are all in place. What's left is choosing effectively—perhaps investigating some alternative route here or there to skirt an obstacle. But when there is no guide for perplexity, when the maps society offers do not conform to experience, choice is a different matter. You are left to your own devices and reason's only resources are first principles. For Gellner, Karl Popper was the paradigm example of an important thinker for whom the maps were still good. After all, he had given one collection of his essays the title *Conjectures and Refutations*. Leading examples of important thinkers for whom the maps were no good were Plato, Descartes, and Marx.

Whether or not someone has denied society's maps was the most important question you could ask, because it framed so deeply what they were about. Without a map there is no fine tuning, no quick check to see if things are moving along the way they should. Your one option is to keep pushing in reason's direction. To suggest that one of these map-less thinkers should have seen "x" or "y" which everyone knows is right there on the map is pointless.[1]

We may borrow Gellner's tale as we turn to Jeremy Bentham and Samuel Taylor Coleridge and ask that crucial question: had they denied the maps of their day?

BENTHAM, COLERIDGE, AND SHADOWS CAST

In 1838 John Stuart Mill wrote an essay on Bentham, a kind of intellectual obituary, and then in 1840 he wrote a second essay on Coleridge, who had died not long after Bentham, in 1834.

The son of James Mill, noted scholar and leading associate of Jeremy Bentham, John Stuart spent several years of his childhood on Bentham's estate. We might rather call it his campus; for by 1817 Bentham had pulled together a set of colleagues, scholars and reformers, including Place, Grote, and Charles Austen, as well as James Mill and his precocious son. John was only twenty-six in 1832, but he had long since played a leading role within the Benthamian set. He had, for example, edited a five-volume work of Bentham manuscripts, *The Rationale of Evidence*, published in 1827, and he regularly contributed to the Benthamian organ, *The Westminster Review*. At the same time, John had become somewhat puzzling and even a disappointment. Place, for instance, wrote in 1838: "I think John Mill has made great progress in becoming a German metaphysical mystic. Eccentricity and absurdity must occasionally be the result."[2]

It is complicated being a disciple. The distance between John and the more orthodox Benthamians was not great, but he was more open to the value of Coleridge and the Coleridgeans than were others. Indeed, his essay on Bentham begins by talking about

Bentham and Coleridge as the two seminal minds of England in their age. In order to capture the essence of either Bentham or Coleridge, one must consider the other. "The influence of Coleridge," Mill wrote, "like that of Bentham, extends far beyond those who share in the peculiarities of his religious or philosophical creed." He then proceeded to capture the differences in these two in a stroke: "By Bentham, beyond all others, men have been led to ask themselves, in regard to any ancient or received opinion, Is it true? And by Coleridge, What is the meaning of it?"[3]

Let's follow Mill's lead, looking first at Bentham and then at Coleridge.

IS IT TRUE?

Bentham is an intriguing figure. He had enormous influence on the law, though he hardly practiced it. He was also influential in politics, but never held office. His interests ranged widely from law and political philosophy to religion and education; though he wrote steadily for many years, most of his writings were never published and those that were, were most often edited by others.

Born in 1748, Bentham went down to Oxford at the age of fifteen. John Stuart Mill tells us that when he was asked to declare his belief in the thirty-nine articles of faith of the Anglican Church, Bentham examined them and expressed some reservations. He was told it was not for boys to question the judgment of the great men of the Church. Bentham signed, but ever after felt it had been immoral; that it had been a lie. Such scruples ran deeply in Bentham.[4]

Not many years later, Bentham studied law and was stunned to discover the custom of making a client pay for three visits to a particular officer of the court, when there had only been one. Again, he was dismayed. After this, he found many more instances of such falsehoods in the practice of law and from these he developed a theory about them, a theory of fictions, a theory that lay at the heart of his work and his influence.

Bentham's work is often characterized in terms of the principle of utility: the notion that the aim of government should be the greatest good for the greatest number. For sure this was a core element in the constructive or synthetic side of his philosophy, but the more important quality of his work was its analytical side and here the theory of fictions was front and center.

Take the central notion of a "right." Bentham observed: "A man is said to have it, to hold it, to possess it, to acquire it, to lose it." As such, we treat it as an object, he continued, "a portion of matter such as a man may take into his hand, keep it for a time and let go again."[5] But this is clearly not the case.

That things are not as they are taken to be—that we do not *have* rights, for example, because they are not the sort of entity one may hold onto—put Bentham in Gellner's mountains without a map. That the notion of a right has been treated in a misleading way opens the door for Bentham to start from scratch. Rights are founded on the perception by individuals that there is a disposition in the way things are that confers a certain benefit. Hence an individual has the right to his personal property, say a coat, only insofar as he is convinced that others would come to his aid should someone try to take his coat away.

What an extraordinary proposal! We are used to the idea of our rights as somehow innate. In the words of the Declaration of Independence: "We hold these truths to be self-evident, that all men are created equal, that they are endowed by their Creator with certain unalienable Rights, that among these are Life, Liberty and the pursuit of Happiness." Bentham's talk of individual perceptions of dispositions in the state of things is in another world. For Bentham, rights are not a gift; we have not been granted them, and we do not possess them by virtue of the nature of things, nor are they posited as self-evident truths in the manner of the postulates of a geometric argument. Instead, rights have being in the primary experience of the way things present themselves to individuals.

This profound shift leads directly to a novel proposition: different individuals in the same society might see things so differently, their experiences might have been so different, that what presents itself as a right to one might not be a right for the other. Can we imagine a legal document holding that citizens within this state shall have rights as they judge them to obtain? But first we should ask Bentham's question: Is such a proposition true? Is there any reason to think that this is the case?

Consider what happens when law and civility break down. It might be a city torn by war with the enemy at the gates. Perhaps it is a catastrophe like Hurricane Katrina. But there's a sea change in attitudes. Individuals are no longer certain that the authorities will maintain order. Possession of your coat is no longer guaranteed. Sadly, we needn't go so far afield. Isn't this exactly what gangs are about? In settings where law and order is less settled, aren't gangs ad hoc structures that step forward to provide that disposition in the state of things that effectively means you have rights?[6]

Bentham saw rights as "fruits of the law, and of the law alone . . . no natural rights, nor rights of man, anterior or superior to those created by the laws."[7] There is an intriguing passage offered a century later by one who saw himself as a progressive reformer just like the Benthamians. It's from Clarence Darrow who also finds the anchor of a central political construct in perceptions of a disposition in the state of things. "Every human being's life in this world," he said, "is inevitably mixed with every other life and, no matter what laws we pass, no matter what precautions we take, unless the people we meet are kindly and decent and humane and liberty-loving, then there is no liberty. Freedom comes from human beings, rather than from laws and institutions." Though Darrow seems to contradict Bentham here, his argument is the same. We are only as free and have rights only so far as our experience in society warrants.[8]

Bentham's analysis of rights denied the leading view of his day, the view of Paine and Jefferson, Burke and Blackstone. This rejection of the natural rights of man was not a rejection of the cause of revolution in America or France. Bentham was among the first foreigners given citizenship in the new French republic. It was instead a fundamental expression of his approach to things.

Bentham applied his theory of fictions by carefully examining propositions because he wished to gauge what was really being said, and then to determine whether it was actually the case. In short, he wanted to know if they were true. It's what you do if the maps are no good.

There is a powerful passage which puts passion to Bentham's sensibilities, but it's not from Bentham himself. And, as it comes from what we might think an unlikely source, it is all the more striking. It is from Thoreau's *Walden*. Thoreau tells us he went to the woods because he wished "to live deliberately, to front only the essential facts of life, and see if I could not learn what it had to teach." And so he entreats the reader to join him: "Let us settle ourselves, and work and wedge our feet downward through the mud and slush of opinion, and prejudice, and tradition, and delusion, and appearance, that alluvion which covers the globe, through Paris and London, through New York and Boston and Concord, through church and state, through poetry and philosophy and religion, till we come to a hard bottom and rocks in place, which we can call *reality*, and say, This is, and no mistake." Through the mud and slush of opinion and delusion which covers church and state, priest and lawyer, you work your way to the essential facts, to what is, and on such rocks, Thoreau adds, "you might found a wall or a state, or set a lamp-post safely."[9] Bentham did not write with such flare, but for sure he gave his life to mud-piercing and foundation-finding.

There is everywhere and always Thoreau's mud. For Bentham, the way through the mud was to avoid deception, and the key was to carefully track fictions. Fictitious entities are treated as if they were real, but they are not. Some, as it happens, are unavoidable. They enter our language once we go beyond simple declarative sentences about things we are pointing at. Things get complicated when we start to generalize, speaking of geese, for instance, rather than the individual goose before us. There is no such thing as "geese." Then there are the metaphors that we've lost sight of as metaphor; as when we speak of "having something on our mind." We are treating the mind as a table, a surface upon which a thought may rest, though we have no reason to suppose it's an object of this sort at all. Such fictions could be avoided, but they are mostly harmless. The metaphors and poetic fancies of the storyteller are essentially idle, from the Homeric "rosy-fingered dawn" to Dylan's answers "blowin' in the wind."

But not all fictions are harmless. And these are the ones that draw Bentham's decided scrutiny and wrath. In a ringing passage, he wrote: "By the priest and the lawyer, in whatsoever shape fiction has been employed, it has had for its object or effect, or both, to deceive, and by deception to govern, and by governing to promote the interest, real or supposed, of the party addressing, at the expense of the party addressed."[10]

What a stunning passage—to deceive and by deception govern.

Paraphrasis

Bentham was dedicated to cutting through these more political fictions, to get at what they really amounted to. He would systematically translate or paraphrase, replacing abstractions with processes and objects. Having established that society and the law are the source of any rights, paraphrasis becomes critical because the law can be shaped and reshaped to meet notions of what is just—but only if we see through its deceptions.

Such careful scrutiny was painstaking and even more tedious to read than to compose. You may recall Sydney Smith's review of Bentham's *The Handbook of Political Fallacies*, where he praises the work of middle-men in bringing Bentham's work to the public. Smith writes: "Mr. Bentham is long; Mr. Bentham is occasionally involved and obscure; Mr. Bentham invents new and alarming expressions; Mr. Bentham loves division and subdivision—and he loves method itself, more than its consequences." Smith did not disapprove of Bentham's views, only his style. Hence the special service provided by editors and commentators who have "washed, trimmed, shaved, and forced [Bentham] into clean linen."[11]

But there was substance to Bentham's belabored style. It was not by accident that he wrote this way. Bentham had a profound distrust of language. It was a most imprecise tool and the issues he addressed required precision. In the absence of clarity the cause of what is right and just has often been perverted—to deceive and by deception govern. Whatever the length, convolution, or obscurity from neologism or tortured phrasing, it was all in the service of clarity.

This matter of distrust of language itself is not unique to Bentham. Consider the writers a good century later, notably e e Cummings, John Dos Passos, and Ernest Hemingway. How sharply their writings reject the style of those who had written in the days before. Henry James, to pick a favorite, invites the reader to sit in a comfortable chair in a warm room dressed in a soft light and join him in reflection on the uncertainties of manner and meaning, and to savor what is too often lost. James's prose plays with clarity. It twists and stutters with asides, contra-positives, and a host of possibilities before finishing a thought. Complex musings served by complex constructions, dressed in a string of clauses and semicolons.

How different are the pages of Cummings, Dos Passos, and Hemingway. They do not invite the reader to get comfortable. They are everywhere edgy, cautioning us to be on guard. There is a profound distrust of language and grandiloquence. Hemingway wrote short declarative sentences: subject, verb, object. Cummings famously abandons grammar altogether, giving us phrases one two three four five just like that. For Dos Passos it is much the same, delivered across whole novels. A favorite, *1919* and in particular its closing section, "The Body of an American," is delivered in an array of voices and fonts. There is the clipped cadence of a government decree delivered by a military spokesman. There is the lyrical eloquence of a feature, perhaps from the *Washington Post* or the *New Yorker*. Then there are the narrator's fragments, tumbling out like change from a vending machine. And lastly, there are the occasional lines from the universal soldier, the everyman who would lie buried in the Tomb of the Unknown Soldier. It is a spectacular piece, and the disdain for those who would gloss over the miserable realities of war with eloquent phrases and carefully parsed complexity is visceral.

Fair enough; I simply thought to share the echo in such writings to Bentham's own distrust of language. His style was not an independent gloss. It was of the very essence.

A Call for Reform

To deceive and by deception govern. Consider an intriguing essay by Douglas Hay. The law might present itself as safeguarding your property, but that is only a mask for a deeper purpose: maintaining the present order and its ruling elite.

Hay's essay is on crime and the law across the eighteenth century. In the early decades of that century, English law counted a small number of crimes as capital crimes, crimes for which the punishment was death, and few souls were executed. Across the century, however, the number of capital crimes increased to a remarkable degree. Where hardly any but the most egregious crimes, like murder and treason, had been punishable by death, now many were. One might face the death penalty for poaching, or smuggling, for robbery, even for blackening one's face. Yet—and this is most stunning—the actual number of people executed did not increase proportionally. Indeed, it continued to be a very small number.[12]

At first glance we might suppose this harsh regime had so intimidated the populace that they avoided committing crimes even of a minor sort, such as killing a rabbit on the lord's estate. But not so. The number of people arrested and charged with these crimes grew with the population. The threat of capital punishment did not deter the criminal.

This is an intriguing problem. A host of "ordinary" crimes are made punishable by death. If the death rate does *not* rise, it seems reasonable to assume it is because the prospective criminal thinks better of it, deterred by the enormity of the cost if he is caught. But then we discover that the number of crimes did not go down at all. In fact, as the population rose across the eighteenth century and as the category of capital crimes increased, the number of people charged and convicted of capital crimes increased accordingly.

When you stop to think about it, perhaps this is not so surprising. Crimes like poaching were not sport, but represented the efforts of people to cope with the scarcity and hardship characteristic of their place in society. Life was hard. They were on the edge.

The question remains: How is it that the number of executions remained so small? The key is that having been convicted of a capital crime, there was a way you could escape execution. A character witness could come before the court and plead for mercy, urging that this episode was an aberration. The convicted felon is really a fine soul, responsible and dutiful, but because of special circumstances, stress, family discord, whatever, had wandered from his or her accustomed, virtuous path. On the strength of this plea the court could suspend sentence. But whose judgment would carry sufficient weight in the eyes of the court? A mother? Father? Friend? Hardly. The court turned to the lord of the manor.

Before the Industrial Revolution, the economic life of England centered on the estates of the landed elite. Everyone, or almost so, worked directly or indirectly for the lord of the manor as farmer, blacksmith, baker, or candlestick-maker.

The real purpose for expanding the number of capital crimes, Hay argues, was to root deference into the everyday behavior of the people. It would have been clear to all that they would quite possibly one day be caught and convicted of a crime punishable by death. Their one way of survival was to make clear to the lord of the manor how very good they were, so that on that day he would speak in their behalf. And so they smiled, and they did as they were told. They knew their place, and they knew it rested on pleasing the lord. They were not free to answer as they might. As Douglas Hay puts it as he closes his essay: eighteenth-century England "was a society with a bloody penal code, an astute ruling class who manipulated it to their advantage, and a people schooled in the lessons of Justice, Terror and Mercy. The benevolence of rich men to poor, and all the ramifications of patronage, were upheld by the sanction of the gallows and the rhetoric of the death sentence."[13]

Hay's notion of the link between law, governance, and deference is not limited to the eighteenth century in England. The very language we use suggests deep resonances. Take the word "civilization." The root of "civilization" has also given us such words as "civil" and "civility." How is it that we have come to capture the might of ancient Rome or of Victorian Britain with a notion of polite behavior?

These words derive from the Latin word, "*civitas*," which means city. To be civil was to behave in a way appropriate for life within the city—or perhaps more aptly, the city brought forth a new quality of behavior, enabling us to become civilized. The city is, perhaps, too much with us for us to appreciate what it meant to the human condition. Aristotle, who viewed man as by nature social, suggested that one must live within the city to be most fully human. The Greeks, by the way, saw the same link between the city and civilized behavior. The Greek for city is "*polis*," from which we derive "polite," as well as "polity" and "politics."

The legal framework that Douglas Hay traces collapsed not through the violence of political revolution, but through profound changes in economic patterns. With the Industrial Revolution workers became unknown figures, anonymous operatives within massive halls of industry. Now the lord of the manor could no longer speak in behalf of those who worked for him. This distorted the purpose of the law which had not been to actually execute wrong-doers, and late eighteenth- and early nineteenth-century judges found themselves with the awful task of choosing between the law and justice. If they convicted a soul for a petty crime the punishment would be wholly disproportionate, but to refuse to convict so as to not have to execute, in the face of clear evidence, was to deny the law itself. From this crisis came a call from the bench for a thoroughgoing reform of the law.

Yet the reforms of the law which took place went far beyond patching up the disproportion of capital punishment and exile. By the early years of the nineteenth century Bentham had devised a whole new approach to the law, an approach that was spare in method and rich in precision. It was the foundation for a radical reconfiguration of the law.

Bentham on the Traditions of the English Legal System

In 1828, Henry Brougham put forward a legal reform bill in the House of Commons. The close of his speech introducing this legislation captures the passion and moral underpinnings of these matters:

> It was the boast of Augustus . . . that he found Rome of brick, and left it of marble. But how much nobler will be the sovereign's boast when he shall have it to say, that he found law dear, and left it cheap; found it a sealed book—left it a living letter; found it the patrimony of the rich—left it the inheritance of the poor; found it the

two-edged sword of craft and oppression—left it the staff of honesty and the shield of innocence![14]

Later, an article from 1851 on recent reforms of England's laws paints a vivid picture of the common law early in the century as "entrenched behind the interests, and defended by the prejudices of the most powerful classes of the kingdom." Yet to question its authority or doubt its wisdom had been deemed "an atrocious libel upon the characters and memories of its architects." The author goes on to describe the unwieldy body of statutes and precedents, the confusion this caused, and the absurdity of expecting any-one, let alone the ordinary layman, to know what the law forbid:

> Though its precepts were to be found only in camel-loads of statutes, many of which had grown mouldy with time, or in adjudications scattered through volumes to be counted by thousands; and though many of the statutes and adjudications had outlived the causes which originated them, and the reasons on which they were based, while others merely tended to increase the general confusion by the war of contradiction they waged among themselves; yet, the most ignorant layman in the kingdom was presumed to know the law and was fined, bankrupted, imprisoned or hung, for not obeying what it puzzled the most acute lawyers to comprehend, and what, after hearing arguments on all sides, it required learned judges months of research and reflection to ascertain; and even then, their *dictum* was law solely because they had declared it so to be![15]

This portrait echoes Bentham's own views in his first work, published in 1780, *Comment on the Commentaries*, a close reading of Blackstone's *Commentaries*.[16] In one telling section he examines the notion of "immemorial usage." He takes each of Blackstone's several examples and traces them to their emergence in a definite time period. Common law is effectively judge-made law, the arbitrary despotism of indi-viduals making the law as they speak. This is Bentham's theory of fictions again. Once we discard the significance of the traditions of the court out of time immemorial, we are left with an individual appointed by and answering to the ruling elite.

For Bentham, the entire foundation of such a legal process was corrupt, and the only way to establish just courts was to make everything explicit. Bentham proposed a rational system, a code of law, assigning a range of punishments which would fit the crime. The judge's responsibilities were transformed into that system with which we are most familiar now—essentially to establish which laws are relevant, whether the accused had broken them, and the sentence within the proscribed range which is best due.

Across the first half of the nineteenth century many reforms in keeping with Bentham's notions were enacted: "Not a single reform has been recently effected in the English law, but its germ may be found in his works, while some are almost literal transcripts of his writings."[17] That Bentham's sweeping new conception of the law was put into place reminds us how many claims there were on change at this time, and how many needs there were to be met.

Bentham and Going for Gold

The nature of the influence of ideas is often ambiguous, witnessed in a harmony that might resonate across just a few bars of a score, but there is no reason to doubt the richness of harmonies in the work of the Benthamian circle. Bentham seems to have been a man of gentle and warm sensibilities, capable of forming lasting friendships and earning the regard and devotion of many. He nurtured the voice of those around him, handing over manuscripts to the likes of Francis Place, John Stuart Mill, and George Grote for them to develop.

If we go back now to that pivotal move by Francis Place to shift the agitation for reform from a call to arms to a run on gold, we can see in it a distinctly Benthamian character. It was a matter of predispositions, of tendencies. Place breathed deeply the atmosphere of the Benthamian circle, especially as he worked on the many jottings and manuscripts of Bentham himself. Here he would have dwelled in the house of Bentham and its focus on the sins of deception. Fictions and fallacies placed a premium on getting down to what was really at issue.

The opponents of parliamentary reform may have spoken of the ancient constitution and of the wisdom of practices out of time immemorial, but what were they really defending? Here is a passage from the closing section of the *Handbook of Political Fallacies*, 1824: "In every political community the holders of the supreme power will, on every occasion in which competition arises, sacrifice the interest of the many to their own particular interest."[18] It was clear to Place that the Duke of Wellington and the members of the House of Lords had done just this, placing their own interests above those of the community, but Place saw further that these interests were not simply a matter of maintaining the status quo and the authority that came with it. Something lay behind the value of this authority: economic well-being. Place challenged the Duke's confidence that he could actually control the economy without regard to the disenfranchised. The run on the bank proved this control was vulnerable. The criticism of reform dissolved.

Of Identical and Representative Species

Going for the gold was not the only instance of the influence of Bentham's theory of fictions. We may recall Grote's pamphlet on reform where he charged that the notion that Parliament was elected by the people was a fiction because the electorate was such a small fraction of the people. A fuller measure of Bentham's influence is found in the core of Grote's study of the earliest chapters of ancient Greek history. The only reliable fact the ancient legends offer critical scholarship is that these tales so freely mix the natural and the supernatural that any effort to separate fact from fiction was little more than guesswork. Whatever historical truth they may contain was beyond our ken. They are of interest, however, because they provoked later commentators who, over time, became the first to separate the swirl of metaphor, imagination, and matters of experience typical of ancient storytelling into the distinct categories of fiction and nonfiction. Grote traced the emergence of historical criticism with commentators increasingly coming to take the measure of ancient claims by how well they conformed to present experience—as when Herodotus wondered how Hercules could have slain a thousand men with the jawbone of an ass, or when Plutarch offered that Hercules must have been a gifted physician. How else would one defeat Thanatos?

This brilliant shift from the history myths might contain to the commentary they provoked was not only the bold print of this section of his *History*. It also defined the thrust of the overall work. The ongoing appeal of the writings of the ancients, of Plato and Aristotle, of Herodotus and Thucydides, is that we can feel their excitement at their discoveries—the new domains reason had led them to.

A key element of Grote's analysis was the open credulity of the ancients. These tales had been taken as literal truths, not as allegories or mere metaphor. To demonstrate this Grote offered a broad comparative study of early storytelling, seeking to establish the open credulity of the ancients by showing a corresponding credulity in, for example, the modern Hindu as attested to by the memoirs of Col. Sleeman; where, for example, he had written: "The Hindoo religion reposes upon an entire prostration of the mind,—that continual and habitual surrender of the reasoning faculties, which we are accustomed to make occasionally, while engaged at the theatre, or in the perusal of works of fiction."[19]

Drawing upon the memoirs of recent travelers was a way of using present facts to shed light on the distant past. The modern Hindu shows us what the ancient Greek thought of his mythology. The most reliable facts of sound history lay in what we can see for ourselves, not what distant reports would have us imagine. Our imaginations are always outclassed by reality.

This brings us to Lyell's *Principles*. Lyell, like Grote, rejected as speculative that history which imagines what causes might have acted in the past, rather than take their measure today as the standard to be applied. The *Principles* opens with an extensive history of geological thought and its often fruitless conjectures and theories, save for those anchored in the proportions of change exhibited in the modern world. Like Grote, Lyell saw the rising use of present facts as the key. Mankind's progress had been a progress of criticism over credulity, of present experience over superstition.

The key to sound history was finding a way to dispel these deceptions by establishing a connection between the evidence and a pointable reality. Lyell's systematic analysis of the present course of nature was just this. To go back to Bentham's theory of fictions, Lyell sought to paraphrase the rock record in terms of processes and operations which could be observed directly.

In an important paper about a dozen years after 1832, the British naturalist Edward Forbes distinguished between representative and identical forms. He was examining the plant and animal life of Great Britain and its connections to the continent. Representative species were analogous to one another in form and function within a community. They were products of the essential dynamics of life, and when found in widely separated regions, they reflected the molding forces of an environment— the power of context. Identical species, on the other hand, when found in different regions, implied that a migration had taken place. Rather than the playing out of the essential laws of nature, they reflected accidents of ancient geography where particular connections had obtained that allowed for their diffusion.[20] We may apply this same distinction when looking at influence, at shadows cast. At times, as with the direct connection between Bentham and Grote, the influence, the shaping forces, were matters of direct connection—the accidents, as it were, of intellectual geography. But in the more expanded view that takes in Lyell's work as well, we may rightly see them as representative species and refer them to deeper molding forces of an intellectual environment: the power of context.

WHAT DOES IT MEAN?

And thus spake on that ancient man,
The bright-eyed mariner.
"The ship was cheered, the harbor cleared,
Merrily did we drop
Below the kirk, below the hill,
Below the lighthouse top. . . ."[21]

And so the ancient mariner begins his haunting tale of how his ship was driven by violent storms to the frigid seas of the South Pole. "The Rime of the Ancient Mariner" appeared in *Lyrical Ballads* in 1798, along with other poems by Samuel Taylor Coleridge and William Wordsworth. This is the more familiar face of Coleridge, the romantic poet of exotic lands and images. But his muse also led him to more prosaic places.

Bentham had found there was no cumulative judgment in the mists of common law, but discrete acts of judgment issued by individuals. This system was open to abuse because the law did not speak directly, but rather indirectly through the judgment of individuals—individuals moreover linked to the ruling classes. His alternative was a legal code, delineating crimes and their punishments; thereby sharply reducing the prerogative of the judge. Now the accused would stand before the court aware of the crime he was being charged with and of the possible consequences.

Analytical principle turned into policy.

Let us seek the same regarding Coleridge and the meaning of things, looking at his last major work, *On the Constitution of the Church and State, according to the idea of each*. The title announces a particular approach, but what does "according to the idea of each" mean?

A True Story from 1982 Not 1832

A long time ago I taught geometry. We used a traditional text following from axioms and postulates through the many theorems which have been taught since Euclid pulled them together over two thousand years ago. At one point we came to the following theorem: that the angle bisector of the vertex angle of an isosceles triangle is also a median. The language here is delightful. An isosceles triangle has two sides of equal length, the vertex angle is the angle between these two sides, the angle bisector a line which cuts an angle in half, and a median is a line which divides another in half. I asked if anyone would like to step to the board to try their hand at proving the theorem.

Stephen volunteered. I had drawn the appropriate triangle. As he walked to the front of the room, Stephen picked up a meter stick. He then proceeded to measure the two line segments of the base on either side of the angle bisector. One was longer than the other. The theorem was false!

It was a lovely moment. Stephen was right, and he was also dead wrong. In fact, he went on to prove the theorem (in that phrase far too common in math and science texts; the proof is left to the reader). What followed was a discussion of the two sorts of triangles in geometry, the ones you draw and the ones you imagine. For sure my drawing was a poor approximation of an isosceles triangle. But if I had been more careful the problem would have remained. No rendering would prove the case. A drawing was not a "real" triangle. It couldn't be. Unlike the lines of experience, lines in geometry have only one dimension. The purpose of the drawing was solely to prompt our thinking. The truths of geometry rest upon triangles of the imagination. That is, reality lies in the world of ideas.

The word "idea" comes from the Greek for form or shape, an unlikely root for a word about concepts and the product of careful thinking. The connection lies in illustrations that Plato would offer again and again: illustrations from the truths of geometry and geometrical reasoning, truths one arrives at by reasoning rather than by measuring experience.

Not to be outdone, the sciences also shifted reality from experience to the imagination. In 1928 Arthur Eddington, an astronomer, wrote a lovely piece about the two tables in front of him as he began to write. One table was a substantive piece of furniture with various tangible qualities like color and texture. The other table was mostly empty space between an array of atoms of carbon and hydrogen and so on that had neither color nor much substance. He then added: "I need not tell you that modern physics has by delicate test and remorseless logic assured me that my second scientific table is the only one which is really there—wherever 'there' may be."[22]

To this we may add that while scientific tables had become quite baroque in their complexity by the 1920s; it was an old table. Democritus, a contemporary of Socrates and Plato, and one of the first to offer that the world was composed of atoms, offered this observation thousands of years ago: "By convention color, by convention sweet, by convention bitter; in reality nothing but atoms and the void."[23] The world is not the way it presents itself.

Back to Coleridge

Common sense would have us distinguish between the real world and those things we only imagine, but mathematics and the sciences have long taught exactly the opposite. This is precisely the observation Coleridge takes off from in his study of Church and State. He tells us: "By an idea I mean (in this instance) that conception of a thing, which is not abstracted from any particular state, form, or mode, in which the thing may happen to exist at this or at that time; nor yet generalized from any number or succession of such forms or modes; but which is given by the knowledge of its ultimate aim."[24]

For Plato and for Coleridge, the way through the thicket is to look through or beyond experience, as geometry teaches us, and find the true form of things with the mind's eye. Don't look for the idea of church or state in the various forms they have taken across the ages. Instead consider their aim, their purpose; what people have been addressing as they worked to create more perfect expressions. These are the postulates of political geometry, the foundational ideas for subsequent theorems.

Bentham sought what things really are by looking straight at them. Judges were individuals and their pronouncements were "judge-made-law," not expressions emerging from the mists of legal tradition. Coleridge takes the opposite tack. In order to genuinely understand institutions like the church or the law that present themselves to us more or less directly, we need to appreciate the idea they instance. It is the idea that is the genuine reality. The particular expression, be it triangle, religion, or state is inherently inadequate, fraught with a host of particulars that are accidents of circumstance, extraneous and distracting; mere prompts to reason.[25] Both Bentham and Coleridge deny the authority of the way things present themselves. Where Bentham denied fictions in favor of pointable reality, Coleridge denied pointable experience in favor of an underlying reality. For the Benthamian this is metaphysical mysticism, but for the Coleridgean an idea is not another word for "fancy, something unreal," but is "the most real of all realities, and of all operative powers the most actual."[26]

So let us now look briefly at Coleridge's "geometrical" study, focusing on what he has to say about the idea which informs a national church.

The Idea of a National Church

The meaning of a national church was not a theological matter, nor was it a constitutional question, a matter of the relationship between the King and the Archbishop of Canterbury. Nor is its ultimate aim the worship of God or the performance of religious ritual. Its purpose, its *idea*, is to support culture. More fully, it is to lead in the cultivation of reflection upon the human condition: the humanities broadly, including both the physical and the moral sciences. The members of this national church, the clerisy, were scholars and teachers of law, medicine, music, architecture, and all the other elements of the liberal arts and sciences, as well as clerics.[27]

How stunning. Coleridge's geometric gaze has led him to a profound underlying aim or motive push in the scheme of things. His analysis is essentially historical. We discern underlying aims in the play of changes, in the growth of institutions and the extension of practices over time. It is, as well, organic. The nation as a whole is an organism, Coleridge tells us, and a moral entity. And the history of a people witnesses the push of its fundamental ideas against the elements of a given era, like the push of a seedling as it asserts its place in forest or field.

Coleridge first tells us of a common ancient practice setting aside a portion of the spoils of war for the community as a whole. This "nationality," as he calls it, is a clear tell that the state has long felt responsibilities beyond the protection and production of individual wealth and property. History shows that the mission of this nationality grows over time in different settings to support collective cultural bodies and practices.

He offers a swirl of the many ways the clerisy have satisfied this purpose across the ages. The Israelites had a system for the support of their priesthood, and Saxon chieftains had a similar system of their own. In time, such systems expanded to include institutions like universities and from there the whole sweep of university study as it has evolved from the quadrivium and trivium of medieval days to the modern university with its array of disciplines: the natural sciences, politics, and the moral sciences, as well as professional studies such as medicine, law, and theology. What is more, in Coleridge's day people had come to see education as so important that there were proposals for a system of national schools for all children. Indeed, Grote, as an MP, seconded a motion for a national school system.

Whether we think of the clan of the Homerids reciting the epics of classical antiquity or the wandering bards of the Middle Ages, or we recall the baths of ancient Rome, the coffee houses of Sam Johnson's England, the Chautauquas of nineteenth-century America, or the blogosphere of our own day, there is a need for an institutional space for talk and reflection, for debate and deliberation.[28]

The clerisy, its structures and institutions over the ages, have responded to an existential need within society. Coleridge approached society as an organism, with a rather Cuvierian regard for structure and function. These various structures satisfying the same basic function were seen by Coleridge as analogous to the relationships between the spiracula of insects, the gills of fish, and the lungs of land animals: different takes toward satisfying the same function.[29] And so, the "proof" within his geometrical study of social institutions was history, history was comparative anatomy, and comparative anatomy is a *map* linking structures to function, a framework for making sense of all sorts of ancient forms and practices.

Stepping back, there is a brilliance to Bentham's insistence that a judge's pronouncement was not the disembodied voice of the court out of time immemorial, but rather judge-made-law—despite the flowery prose of Coke and Blackstone to the contrary. There is a matching brilliance to Coleridge's insistence that parish church and grand cathedral are not the Church of England, nor is it the thirty-nine articles of the Anglican Creed, or any other theological nicety. It is, instead, the cultivation of the understanding wherever that occurs.

Analytical principles turned to policy.

John Stuart Mill agreed. He saw Coleridge's view of a national church as a powerful vision, not least because it undermined the complacency of the Church of England. Here was a new face to conservatism, one that profoundly challenged the way things are. Benthamians, Mill urged, should see Coleridge and his disciples as kindred spirits, equally discomfiting to the complacent. The two "are opposite poles of one great force of progression . . . Each ought to hail with rejoicing the advent of the other."[30] Here, no doubt, is Mill wearing that mystical coat of Coleridgean splendor that made Francis Place uncomfortable.

Of Mirrors and Lamps

Coleridge was a Romantic poet. Is his Romanticism best seen as a literary style, only relevant to his poetry? Or was it something deeper; something essential that gave form to his analytical writings, as well?

As with Bentham, whose style reflected the essence of his understanding, so also with Coleridge. We lean heavily here on the central theme of M. H. Abrams's rich study of late eighteenth- and early nineteenth-century English letters, *The Mirror and the Lamp*. Abrams argues that there had been a profound break at the turn into the nineteenth century, a break in the nature of poetry. The classical tradition had held poetry to be an imitative art, and had derived critical principles from this notion. Romantic theory, exemplified in Wordsworth's Preface to *Lyrical Ballads* and Coleridge's *Biographia Literaria*, conceived of poetry as the expression of the inner life of the poet.

Abrams tells us that the "theorist who held art reflected nature was committed to looking 'out there,' rather than into the artist, for the subject matter of a work." This classical theory raised such issues as whether poetic images should be of actual reality, or of a deeper, more essential and refined reality. This was a complex matter, spilling over into perceptions of what is natural and what artificial. It was also directly relevant to how history and poetry were distinguished. "It had been common since antiquity to oppose history and poetry," Abrams writes, "and to base this distinction on the ground that poetry imitates some form of the universal or ideal instead of the actual event."[31]

We can see how this fits in with Mitford's understanding of the earliest accounts and legends of ancient Greece. Their historical character had been corrupted and was only safe with regard to the nonuniversal, nonideal elements—in other words the nonpoetic elements, which would be the discrete particulars of dates, events, colonial migrations, and military expeditions.

In contrast, the romantic theory held the content of poetry to be internal, shaped by the "forces inherent in the emotions, the desires, and the evolving imaginative process of the artist himself."[32] The poem was not about what was "out there"; rather, it was about the feelings of the poet. To take a classic example, Wordsworth's "I wandered lonely as a cloud" is less about the scenery than about how it makes him feel. For the classical theory, the poet was a mirror; for the romantic, the poet was a lamp.

With Coleridge's study of church and state there is a corresponding shift from the outward shape of things to their inner vitality. Church and state can only be understood if we see them as expressions of the idea which informs or shapes them, a process of more fully coming to be over time. In this passage Coleridge ties the historical development of English history to the "gradual realization" of the idea of the English constitution:

> Our whole history from Alfred onwards demonstrates the continued influence of such an idea, or ultimate aim, on the minds of our fore-fathers, in their characters and functions as public men, alike in what they resisted and in what they claimed; in the institutions and forms of polity which they established, and with regard to those, against which they more or less successfully contended.[33]

History is the coming to be of the prevailing idea. Coleridge's nonfiction is not Romantic because *he* is the lamp, the source of light and motion, rather the ideas are the lamp. What is compelling in both of these faces of Romanticism is the push from within: the sensibilities that inform the motion of the individual who wanders lonely as a cloud has its analogue in the unfolding idea which transforms the kingdom of Alfred the Great into the parliamentary monarchy of Coleridge's day.

Coleridge's notion of the clerisy, we may add, reflects the rise of the sciences and their many-splendored theories, a rise marked by the striking success of the British Association. The first meeting of the Association had been at York. The second meeting was at Cambridge. Coleridge attended. He had for several years been living with Dr. John Gillman. A victim of various ailments and heroin addiction, Coleridge was by now quite frail. Nevertheless, he relished the vitality of various talks and issues at the meeting, engaging in a variety of matters over three days. It was one of his last excursions. Bentham had died the previous year. There was little left but their shadows.

Shadows . . . there is a lovely line, from Emerson, that an institution is the shadow cast by a single man. We have considered the shadows cast by Bentham; let us now seek the measure of those cast by Coleridge.

On the Measure of a Shadow

For Thirlwall, myths represented the efforts of early humankind to explain how the world had come to be the way it was, providing both a physical and a moral cosmogony—both why the world was the way it is and why we should act the way we do. This was clear in his read of the many legends surrounding Hercules. He sorted them into two batches. Those of Hercules the greater addressed the concerns of those within the earliest communities. How had fields come to be cleared of boulders and wild animals? How had the river been led to flow where it was so useful? These were the great labors of Hercules. There were, as well, the deeds of the lesser Hercules. Maintaining the cause of the weak against the strong, the innocent against the oppressor, punishing wrong and subduing tyrants, this Hercules established the proportions of propriety for a later era, addressing the concerns of a more settled society.

Within an oral culture the community as a whole generates its mythology. Certain deeds had been accomplished; traditions and notions of propriety had been set. Where did they come from? Why are they the case? The answer lay in the mighty arm of Hercules. Like kinship and migrations, the hero was a conceptual device. Further, the mythological genius was dynamic. Myths are a living form, responding creatively to the changing concerns and needs of the community.

As we saw in chapter 3, Thirlwall tested this hypothesis, considering two other ancient heroes, Theseus and Minos. In both cases there is a like development, where the hero is first a mighty figure, wrestling animals or conquering enemies, and then there are tales of a less fabulous aspect, as our hero rights wrongs or enacts political reforms.

These myths were not historical reports enlarged upon by wandering bards. They were the voice of a people making sense of their world, put forward, preserved, modified, and extended to maintain their vitality: an underlying idea continually pushing itself against prevailing conditions. They were living forms and insofar as conditions approached one another, so too did the correspondence in form. The exploits of Hercules the greater, the young Theseus, and the "Phoenician" Minos present variations within a type, corresponding to the first settlement of community. Within their different settings, these heroes cleared land and sea of marauders and monsters, and so enabled the establishment of community. Shifting to a later day, with Hercules the lesser, Theseus the wise ruler, and Minos the law-giver, we find a common variation under a new aspect. What is this but a version of Sedgwick's Eastern Alps thesis! Thirlwall has recognized the true correlation between separated deposits, a correlation that hinged on the interaction between the mind of the storyteller and the concerns of his audience.

Here is the fullness of Thirlwall's alternative to Mitford's "stratigraphical" history. He has presented a developmental, "ecological" history where the essential conceptual stock of the ancient imagination is played and replayed against changing conditions—the continual adaptation of the past to meet the demands of present sensibilities. These conceptual devices, collectively and individually, are the idea within myth, an idea representing the force of the imagination as it asserts itself in the effort to make sense of its world.

Here is history as lamp, rather than mirror.

Poetry of Rock and Fossil

What had been fact for Mitford became perception for Thirlwall, a shift that echoes our reading of Coleridge as a Romantic historian. As Romanticism conceived of poetry as the expression of the inner life of the poet, so history became the uncovering of the idea within things that provided its motive push. Thirlwall read myth for the mythological mind, linking the artist of antiquity with the voice of the people and directing interpretation away from the events in the tale and toward the mind of the storyteller.

We may return now to the chapel at Trinity College on that December day in 1832 as Adam Sedgwick examines the true meaning of education in these days of so many changes. Speaking of the sciences, he turns to geology where every new life form in the fossil record demonstrated the active providence of God; in each case, new organs "were exactly suited to the functions of the beings they were given to."[34] The appearance of new life was a mystery, affirming the "creative interference" of God.

There is something crucial here. The process of becoming is beyond our grasp, and so signals the handiwork of God. I am reminded of an essay written by the young Newton where he envisions a vast grid of cells which fill space. Matter occupies some of these cells and not others. At succeeding moments which cells have matter and which do not changes by the hand of God. Reality is a series of stills. There is no passage. Just a series of discrete jumps, much the way a digital screen works. No act of becoming. Simply being.[35]

There is a mystery in the fossil record. Each gap, and most especially those that separated successive epochs, represented a moment of creative activity beyond our knowing. This view that we are unable to understand the origin of new forms underlines that the task before the historian was far less a matter of the processes of passage than evoking given moments or states: exactly the opposite take of Grote and Lyell's work, which were all about processes of change.

Sedgwick amplified these views in the lengthy preface he added to the fifth edition of his *Discourse*. He proposed that we see the history of life as a succession of dominant types. These types had not emerged gradually, but appear abruptly with complexity and variety. We may take as representative his remarks on the earliest fish, which "did not rise up in nature in some degenerate form, as if they were but the transmuted progeny of the *Cephalopoda*; but they started into life (if we are to trust our evidence) in the very highest icthyic type that ever was created."[36]

Each of these types stood at the center of a system: "There was a time when *Cephalopoda* were they were the highest types of animal life. They were then the *Primates* of this world; and corresponding to their office and position, some of them were of noble structure and gigantic size."[37] There was thus a general plan to nature's history, identical across the ages. *Cephalopoda* and *Primates* had spread out within their epochs, filling many ecological stations, adapted in successful ways. History was comparative anatomy. Systems, which had begun as broad assemblages of fossils had evolved into ecological wholes.

There remains but one more step. It is the step that links the abstract form of a system, its plan, the cumulative expression of all the evidence and its many faces and aspects, with the mind of God. Here is the sublimity of early Victorian natural history with what we may take as the credo of her finest spokesman. "What we do believe is," Sedgwick wrote, "that before the creation of all worlds, there was an *archetype* of nature (dead as well as living, past as well as present) in the prescient mind of God."[38]

Sedgwick's archetype fuses natural history and geology. But it does more than this, for the archetype is the pattern by which all systems have unfolded. It is like the genetic code for an entire geological epoch, and again for each succeeding epoch. As such, we can see that the archetype is an analog to a Coleridgean idea. It is the underlying reality within experience. Coleridge had written of Church and State according to the ideas that had informed their development over hundreds of years. In the same way, Sedgwick has written of the systems of life according to the archetype which had guided their development over the vast spans of geological time. Coleridge had seen states and religions at any particular time as fraught with particulars that are accidents of circumstance, extraneous and distracting; mere prompts to reason, the underlying ideas were "the most real of all realities, and of all operative powers the most actual." So, too, for Sedgwick, despite the array of particulars at any locale fraught as it would be with accidents of circumstance associated with that particular ancient landscape and all the succeeding events of that region, the archetype was constant and the most real of all realities and of operative powers the most fundamental, coming as it did from the prescient mind of God. To contemplate this archetype was to reflect on the design which embraces all of nature.

That is quite a map.

NOTES

1. This story is alluded to in Gellner's *Legitimation of Belief*, where a section of the opening chapter ("Get Lost") discusses this experience and proceeds to the central matter of reading a text where the maps were no good.
2. Wallas, *Life of Francis Place*, 91.
3. Mill, "Literary Remains," in *Collected Works*, vol. x, 119.
4. Mill, "Works of Bentham," in *Collected Works*, vol. x, 81.
5. Bentham, *Pannomial*, vol. iii, 218.
6. Sadly, the same dynamic can often be seen in our schools. In some we see the expectation that civility will obtain. In a congested hallway, students will make their way through the jostling crowd with a minimum of conflict. But in far too many of our schools this is not the case. There are many conflicts and it is not unusual for fights to break out. Bentham's discussion of rights suggests we look at what students see in the disposition of things. They feel threatened at every jostling, and even a glance can set things off. While students may put their backpacks down, only rarely can they put down the problems of their homes and neighborhoods. Schools have to work very hard to overcome the lack of security and civility in the streets. Too few succeed.
7. Bentham, "Pannomial," in *Collected Works*, vol. iii, 221.
8. Darrow, "Closing Argument," Recorders Court, Detroit Michigan. Darrow was speaking on behalf of a black man who had just moved into a white neighborhood. In defending himself, his family and friends, and his property against a mob, one man was killed. The accused were acquitted.
9. Thoreau, *Walden*, 127.
10. Bentham, "Fragment on Ontology," *Collected Works of Jeremy Bentham*, vol. viii, 199; see also de Champs, "The Place of Jeremy Bentham's Theory of Fictions in 18th Century Linguistic Thought."
11. Smith, "Bentham on Fallacies," 209.
12. Hay, "Property, Authority, and the Criminal Law," in *Albion's Fatal Tree*, 17–64.
13. *Ibid.*, 62–63.
14. Brougham, "Speech in the House of Commons," *Jurist* ii (1828): 6.
15. Kettel, "Law Reform in England," 35.
16. Bentham's *Comment on the Commentaries* forms the first section of *A Fragment on Government*, vol. i of *Collected Works*. It includes a host of denuding passages on the traditions of common law, such as this one on judge-made law: "He [Blackstone] gives the presence of *one* man at the *making* of the law, as a *reason* why ten thousand others that are to obey it, need know nothing of the matter," vol. i, 233. (Note, the *Commentary* is not included in Bowring's original collection of Bentham's works.)
17. Kettel, "Law Reform in England," 41.
18. Bentham, *Handbook, Collected Works*, vol. ii, 482–83.
19. Sleeman, vol. i, 70, 227 and vol. ii, 51, 97; cited in Grote, *History*, vol. i, 414.
20. Forbes, "On the Connexion," 336.
21. Coleridge, "Rime of the Ancient Mariner," 405.
22. Eddington, *Nature of the Physical World*, xii.
23. De Santillana, *Origins*, 145.
24. Coleridge, On the Constitution, 11.
25. *Ibid.*, 39.
26. *Ibid.*, 18.

27. *Ibid.*, 50, 52.
28. Speaking of the coffee houses of the latter seventeenth century, Macaulay writes: "No Parliament had sat for years. The municipal council of the City had ceased to speak the sense of the citizens. Public meetings, harangues, resolutions, and the rest of the modern machinery of agitation had not yet come into fashion. Nothing resembling the modern newspaper existed. In such circumstances the coffeehouses were the chief organs through which the public opinion of the metropolis vented itself" (*History of England*, vol. i, 128).
29. Coleridge, *On the Constitution*, 21.
30. Mill, "Literary Remains," *Collected Works*, vol. x, 146.
31. Abrams, *The Mirror and the Lamp*, 35, 101.
32. *Ibid.*, 46.
33. Coleridge, *On the Constitution*, 18, 19.
34. Sedgwick, *Discourse*, 20, 23.
35. Newton, "De Gravitatione," in Hall and Boas, *Unpublished Scientific Papers*.
36. Sedgwick, *Discourse*, 5th ed., ccxvi.
37. *Ibid.*, ccxv.
38. *Ibid.*, ccxix.

8

IN THE WILD

The year 1832 finds the young Charles Darwin at sea aboard the HMS *Beagle*. He had finished at Cambridge in the spring of 1831 and was delighted to have landed the position of naturalist on a voyage around the world. The underlying purpose of this and many similar voyages was to take stock of Britain's empire and the logistics of maintaining it. The *Beagle*'s particular mission was to map the coastal waters and inland regions of South America, especially the many channels of the archipelago at the tip of the continent, known ironically as Tierra del Fuego, land of fire.

By this time, Britain had already established a far-flung empire and maintaining her holdings had long been a serious enterprise; one that routinely required rounding the treacherous tip of South America. The *Beagle* would spend many months seeking an effective sheltered passage among the islands of Tierra del Fuego and harbors and channels deep enough for ships of the line.

Darwin was not a regular member of the crew, nor was he a member of the Royal Navy. He had been chosen by Captain Fitz-Roy. We have become accustomed to the notion of a scientist on board a military vessel, especially via science fiction, from Dr. Zharkov with Flash Gordon to Mr. Spock on the Starship Enterprise. But why would Fitz-Roy have reached into his own pockets to have a naturalist for this voyage? The answer in part takes us back a little farther.

JOSEPH BANKS AND THE MAGIC OF TAHITI

As the Renaissance had been captivated by the geographical expeditions of Marco Polo, Columbus, and many others, so the nineteenth century was taken with the great voyages of Humboldt, Darwin, Huxley, and Wallace. These intrepid explorers were armed with thermometers, sketch pads, and collection bottles, not carbines. The riches they sought were data, and they returned with coffers filled with stories of the exotic rather than silver and gold.[1]

The formative event here had been the voyage in 1768 of Captain Cook's *Endeavour* with its young botanist, Joseph Banks. The voyage was part of the Royal Navy's standing mission to locate and secure the resources of Britain's burgeoning empire, especially lands of the south Pacific. As it happened, astronomers had determined there would be a transit of Venus in 1769 and not another one for over a hundred years. There was enticing value to these rare transits; Edmund Halley, building on the work of Johannes Kepler, had worked out early in 1716 that you could calculate the distance from the Earth to the sun by comparing the time it took Venus to pass across the sun as witnessed from two different places on Earth. But it was a subtle effect and the further apart the sightings were, the better the data would be. That's why the *Endeavour* planned to set up shop for observations in the remote Pacific island paradise of Tahiti, more than nine thousand miles from the observatory at Greenwich.

Joseph Banks, however, was not part of the astronomical team funded by both the Royal Society and the Admiralty. He had heard of the expedition and offered to be its botanist, offering as well the princely sum of £10,000 to pay for a team of naturalists, including two artists, and their equipment. The Navy and the Royal Society quickly agreed and so it was that the young Joseph Banks first set eyes on the tropical paradise of Tahiti in April of 1769. Banks's routine included botanical drawing, electrical experiments, and animal dissections. He regularly fished specimens from the sea, shot or netted wild birds, and observed meteorological phenomena, but his interactions with Tahitians quickly absorbed his attention. This was, to a degree, because of the striking beauty and sensuous character of the Tahitian women and the fact that the prevailing mores welcomed sex as natural. His journal often recorded seductive encounters, and as Richard Holmes recounts, Banks "was increasingly prepared to abandon European inhibitions, including his clothes. He noted frequently, 'I lay in the woods last night as I very often did,' by which one can understand he was probably with Otheothea (a Tahitian woman)."[2]

There was more to it than the pleasures of a night in the woods. Banks was virtually the only member of the crew who bothered to learn more than just a few Tahitian words. But he not only learned the language; he also examined the customs and social sensibilities of the Tahitians. This proved to be valuable on several occasions, but nowhere more dramatically so than when the expedition's quadrant was stolen. The quadrant, a critical piece of equipment, had been kept secure on the ship. But as the day of the transit approached, it had been brought ashore, placed in Captain Cook's tent—and promptly stolen.

After all the careful preparation, resources organized, and monies spent, the entire mission was in jeopardy because a Tahitian had been fascinated by the intricacy and beauty of a scientific instrument.

Banks acted swiftly, rousing one or two crew members and setting off to catch the thief. They tracked him for seven miles, running and walking through the tropical heat and into terrain far removed from the protection of the *Endeavour* or the Tahitians who knew him.

What made Banks take this risk? He was not a member of the crew nor was he a part of the astronomical team. He had no formal obligation, yet he threw himself into the mix. Why? No doubt it was an impulsive act, and no doubt there was an arrogance to it, but it was not the arrogance of an aristocrat ruffled by the audacity of a native who dared to interfere with the British Empire. This was not a matter of King and Country. Banks respected the scientific mission and the captain of the *Endeavour*. He acted out of loyalty to science and a friend. And with these, it was the arrogance of youth, the arrogance of an accomplished young man (he was twenty-six at the time) who embraced life with open arms.

As they closed in on the thief, Banks sent one of the midshipman back to call for support. Then, as he approached a set of huts on the crest of a hill, a Tahitian brought out part of the quadrant. Banks recounts in his journal that "many Indians gatherd about us rather rudely." Surrounded by a hostile crowd, Banks following a Tahitian custom he had seen before, marked a circle in the grass and sat quietly down in its center. Without bombast or threats, he explained what had happened and began to negotiate.

How remarkable! Of all the many encounters between European explorers and various indigenous peoples, how often had Europeans quickly resorted to gifts that would amaze, trinkets and such-like, followed by weapons that would intimidate? But here is a young man who marks a circle, sits down, and talks. In time, the quadrant was returned piece by piece. The journal entry continues: "After walking about 2 miles we met Captn Cooke with a party of marines coming after us, all were you may imagine not a little pleasd at the event of our excursion."[3]

Banks's transit across part of the Tahitian terrain was a great success, enabling the subsequent measurement of the transit of Venus across the sun, which was equally successful. The data suggested a distance from the Earth to the sun of over 95 million miles, within 2 percent of the modern value.

But greater still was Banks's transit across Tahitian culture. Upon leaving Tahiti after a three month stay, Banks wrote an essay titled "On the Manners and Customs of the South Sea Islands." It is packed with observations on Tahitian cooking, boat-building, tool-making, fishing, dancing, and ceremonial practices.[4] A more telling piece is one he wrote a couple of years after returning to England, "Thoughts on the Manners of the Otaheite." It is a more reflective essay, one that seeks the deeper differences in European and Tahitian culture. At one point, following an account of the way European women are squeezed by garments with stays "scarce less tenacious than Iron," and other contrivances which altogether conceal their natural form, Banks wrote of Italian painters who "clothd their goddesses and angels in loose folds of Cloth not shapd to their bodies exactly as the Otaheiteans now wear theirs." And then he movingly added: "An Otaheitean on the other hand will by a motion of her dress in a moment lay open an arm and half her breast the next maybe the whole and in another cover herself as close as prudery could contrive and all this with as much innocence and genuine modesty as an English woman can shew her arm."[5]

It is clear from this and other writings that Banks had seen more than cultural differences during his stay with the Otaheite. He had seen a nobility and grace that mocked European culture.

THE NOBLE SAVAGE

Banks had met the noble savage. In terms of the broadest sweep of Western history, there had emerged across medieval Christianity an understanding of humankind's place in nature captured in the metaphor of a great chain of being that reached from the lowest plants and animals through to the heavenly host and God Almighty. At the center were humans. As we had both body and spirit, we linked the world below with the world above. That we were both body and soul also addressed the question of whether we are naturally good. Our divine selves are good, but our bodies are a source of corruption, and life unfolds as a struggle between these two powers. Thus the charge to lead the life of Christ was often translated into the need to rise above the temptations of the flesh.

At the same time, there was a long-standing minority report. This alternative view seized on the innocence of youth. It lent itself to the idea that in olden times, when life was simple, mankind had been innocent and gentler. It is more or less in this vein that we find Jean-Jacques Rousseau's *Discourse on the Origins of Inequality among Men*, which raised the vision of the natural state of humankind as marked by empathy for the suffering of others. Rousseau's *Discourse* had been composed in 1754, several years before the voyage of the *Endeavour*. As the eighteenth century progressed, the minority report fused with the contrast between European culture and the host of indigenous peoples Europe was coming to know.

Innocence was no longer restricted to the distant past. It was also true of native peoples. So, for example, we find Ben Franklin outraged at the savagery of white vigilantes who massacred 140 Indians, chiefly women and children, in 1763. Franklin described two episodes in a pamphlet titled "Narrative of the Late Massacres," on the differences between savages and White Christians. In one episode, Franklin relates, Catawha Indians sued for peace with Mohawks of the Six Nations. The Mohawk council rejected their offer, and added:

> While you are in this Country, blow away all Fear from your Breasts; change the black Streak of Paint on your Cheek for a red One and let your faces shine with Bear's-Grease: You are safer here than if you were at home. The Six Nations will not defile their own Land with the Blood of Men that come unarmed to ask for Peace.

The contrasting episode concerned the recently slaughtered Indians:

> These poor People have always been our Friends. Their Fathers received ours when they were Strangers here, with Kindness and Hospitality. Behold the Return we have made Them!—When we grew more numerous and powerful, they put themselves under our *Protection*. See in the mangled Corpses of the last Remains of the Tribe, how effectually we have afforded it to them![6]

For Franklin, there was no question about who were the savages.

Upon returning to England Banks discovered he had become a celebrity. Newspapers and magazines ran lengthy articles about his exploits. Captain Cook and the voyage of the *Endeavour* captured the collective imagination. Banks and Cook were summoned to meet the King, and this interview led to a long friendship between George III and Banks, leading in turn to the development of Kew Gardens. Banks and his natural history team had collected over a thousand new plants and five hundred animal skins and skeletons, along with a host of artifacts from indigenous peoples. In 1778 Banks became president of the Royal Society, at the strikingly young age of thirty-five. He also became a member of Sam Johnson's Club, along with Sir Charles James Fox. Later, reflecting on his work, Banks wrote that he had been "the first man of scientific education who undertook a voyage of discovery."[7]

Darwin was born in 1809, some forty years after Joseph Banks's voyage to Tahiti. While Banks never succeeded in writing up his experiences in any more extended form, their impact was still great. The voyage of the *Endeavour* left its stamp upon both the popular imagination and the scientific community. In the decades following there were many scientific accounts of native peoples, climes, and cultures. We know the young Darwin was familiar with this literature; his *Journal of Researches*, an account of the voyage of the HMS *Beagle*, refers to many of them, including the voyage of the *Endeavour*, which had stopped briefly in Tierra del Fuego on its way to Tahiti. In a letter to one of his sisters, Darwin would write: "My first intro to the notorious Tierra del F was at Good Success Bay & the master of ceremonies was a gale of wind.—This place was visited by Capt Cook; when ascending the mountains, which caused so many disasters to Mr. Banks, I felt that I was treading on ground, which to me was classic."[8]

We may add one further curious item. As the *Endeavour* left Tahiti they took along an Otaheite, a man named Tupaia, who proved useful both to the captain in the further exploration of the Pacific and to Banks as he began to write up his experiences.[9]

OTHER GUESTS ON THE *BEAGLE*

As it happened, Darwin was not the only guest aboard the *Beagle*. How these others came to their place on board is its own intriguing tale.

Only a few years senior to Darwin, Fitz-Roy had been elevated to the rank of captain at a very young age, in December of 1828, to replace the former captain of the *Beagle*, who had committed suicide. Fitz-Roy acquitted himself very ably, taking firm charge of the ship and her appointed tasks. While he was evidently of a temperament and training well suited for careful measurement, he had many taxing difficulties with the Fuegians of various settlements. They possessed no armies and there were no organized violent confrontations, but they stole what they could with no apologies. We have already met with somewhat similar sensibilities in Tahiti, yet there was a difference. Perhaps, if the Fuegians' early contacts with Europeans had regularly included people like Joseph Banks, who learned their languages and cultural practices things would have turned out differently . . . but life was very different in Tierra del Fuego. These islands are no tropical paradise. They present as harsh and punishing an environment as any on earth. The islands are routinely bitterly cold and subject to strong winds and heavy rains. On top of this, the Fuegians had little technology to help them cope with so severe a setting. They were hunter-gatherers, and their arsenal of weapons consisted chiefly of stones and spears. Here is an observation from Captain Fitz-Roy: "They never attempt to make use of the soil by any kind of culture; seals, birds, fish, and particularly shell-fish, being their principal subsistence; any one place, therefore, soon ceases to supply the wants of even one family; hence they are always migratory."[10] Life was harsh and any advantage, however gained, was valuable.

On that first voyage for Captain Fitz-Roy, The *Beagle* had several rather threatening experiences, including the theft of a long boat, which left half a dozen men stranded on an island, their whereabouts unknown to the main crew. The stranded men were able to fashion a meager raft and two brave souls set out to cross various channels in hopes of finding the mother ship. By good fortune, they were spotted by the *Beagle* and rescued. Fitz-Roy immediately set out to retrieve the other men and to track down the culprits who had stolen the long boat, but to no avail. Along the way, they captured some women and children that had been left behind, thinking they could be held as hostages for the return of the boat and its supplies. But the natives showed no interest in such a trade. After a host of complications, Fitz-Roy found himself with four young Fuegians. Contemplating his predicament, he became enchanted with the idea of taking them back to England, where they could be given all the advantages of Christianity and English civility and then be returned to the islands to lead a civilizing mission.

One of the four died early on, but the remaining three flourished on the trappings of aristocratic life for the year or so they spent in England. Fitz-Roy was a member of the landed elite; his grandfather was the Duke of Grafton. Here was a real-life episode to match *My Fair Lady*. Indeed, the Fuegians absorbed English manners and language so well they had an audience with King William IV and his Queen.

We have little evidence of how these Fuegians felt about what was happening to them. We know they were young: the one known as Jemmy Button was fourteen, Fuegia Basket was but nine, and York Minster was twenty-six. The rest is pretty much left to the imagination, but you would be hard-pressed to come up with an instance of a more dramatic upheaval in cultural and physical settings than the move from a Fuegian hut to an English country manor.[11]

Fitz-Roy's account of the voyages of the *Beagle* is a fascinating document. You can see him struggling to capture the range of his feelings and judgment. He begins by observing that during the initial voyage in 1830 he had time to study his Fuegian companions, finding them "far, very far indeed, were three of the number from deserving to be called savages—even at this early period of their residence among civilized people."[12] Yet he quickly adds a comment on the regular acts of cannibalism of Fuegians, a "most debasing trait" which, he goes on to explain, is often but not always tied to the scarcity of food. This juxtaposition of strengths and most disagreeable habits is captured nicely in an observation the captain offers as he justifies his decision to bring the Fuegians home to England.

> Disagreeable, indeed painful, as is even the mental contemplation of a savage, and unwilling as we may be to consider ourselves even remotely descended from human beings in such a state, the reflection that Cæsar found the Britons painted and clothed in skins, like these Fuegians, cannot fail to augment an interest excited by their childish ignorance of matters familiar to civilized man, and by their healthy, independent state of existence.[13]

What a mix of notions—to have descended from such barbarians is most disagreeable, and yet England had itself been barbaric in a definable past.

But if Fitz-Roy was caught in the middle, how much more so the Fuegians themselves? Late in 1832, the *Beagle* sailed into the archipelago on its return voyage, the voyage Darwin was on, seeking safe harbor to drop the Fuegians off with a young Anglican missionary, Richard Matthews. The Fuegians, who had initially been excited about returning to their homeland, became agitated. York laughed at the natives on the shores, calling them "monkeys—dirty fools—not men."[14] Jemmy assured everyone that they were not like his people who were both good and clean, and Fuegia, who would have been but eleven or twelve by now, was shocked and ashamed.

Fitz-Roy noted this confusion of feelings. He was also struck by a corresponding confusion of languages. On several occasions he notes that Jemmy seemed to have lost his native tongue and was having a hard time shifting back to his native culture. At one point Jemmy confessed to the captain that he knew very little of his own language, and that he mixed words of English and Tekeenica when he talked to his family.[15] It would appear the disconnect between Fuegian and European experiences, language, and culture was so profound that Jemmy could hardly keep hold of both at the same time.

When supplies had been put ashore and a set of huts built for the Fuegians, along with a building and gardens for Matthews and the mission, the *Beagle* continued on with its own mission. They returned after a most anxious week or so to find Matthews deeply disheartened about his prospects, and Fitz-Roy decided to abandon the missionary project. Most of Matthews's goods had been stolen, his life threatened, and Jemmy, Fuegia, and York had disappeared. When, sometime later, Jemmy was spotted, everyone was stunned by the transformation. Here is Fitz-Roy's account, as he sees that Jemmy had detected him:

> A sudden movement of the hand to his head (as a sailor touches his hat) at once told me it was indeed Jemmy Button—but how altered! I could hardly restrain my feelings, and I was not, by any means, the only one so touched by his squalid miserable appearance. He was naked, like his companions, except a bit of skin about his loins; his hair was long and matted, just like theirs; he was wretchedly thin, and his eyes were affected by smoke.[16]

Jemmy Button came aboard, and he and the captain had one last conversation. His two or three years with the *Beagle* and English society was but a thin veneer quickly shed, like a coat as you come in out of the rain. Fitz-Roy quickly wrapped up this episode, explaining away the failure of the missionary scheme as understaffed, and taking some comfort regarding Jemmy in his continued open, good spirits and his earnest gratitude for all that the captain and others had done: "That Jemmy felt sincere gratitude is, I think, proved by his having so carefully preserved two fine otter skins, as I mentioned; by his asking me to carry a bow and quiver full of arrows to the schoolmaster of Walthamstow, with whom he had lived; by his having made two spear-heads expressly for Mr. Darwin; and by the pleasure he showed at seeing us all again."[17]

He concluded that "as nothing more could be done, we took leave of our young friend and his family, every one of whom was loaded with presents, and sailed away from Woollӯa."[18] Such were the complexities of European encounters with the wider world.

CHARLES DARWIN AND BERGERET'S LEMMA

Plant a carrot, get a carrot—not a Brussel sprout,
That's why I like vegetables, you know what they're about.

As these lyrics from the musical "The Fantastics" suggest, nature is remarkably regular. Carrot seeds do routinely grow into carrots. But if carrots routinely produce carrots, if dogs always give birth to puppies, cats to kittens and frogs to tadpoles, then how is it possible that frogs became dogs? Or to put matters in truer proportions: How are we to understand that we were once worms, and before that, along with cats, crayfish, cauliflower, and carrots, we were all single-celled organisms of the sort that forms a scum on a pond in the middle of the summer?

That life has evolved is a cornerstone of the modern sciences. Others before Darwin had argued that life had evolved, but their arguments had been rejected. It was Darwin who transformed the hypothesis into a mainstream scientific theory. He did not deny "plant a carrot, get a carrot." He could not. It is the overwhelming experience of us all. What he did was to examine with great care just what it means to be a carrot (or a frog, or a dog) not by examining the physiology or anatomy of these forms, but by questioning membership in the set. Are all carrots identical? Of course not. How much room is there between the not identical but still very much like the others? And then, what if such small differences could be inherited? That would allow things to change so subtly we would never see it, even though it would be happening all around us. These are the invisible evolutionary bits he posits.

A CANTERBURY TALE

This business of indiscernible change is intriguing. Let's pause to consider an example borrowed from the study of languages, noting as we do a reflection Lyell offered in another letter to Whewell, this one from March of 1863. Lyell writes of his "habit of thinking of the changes of languages when I wanted to clear up some of my own ideas about transmutation."[19] Geoffrey Chaucer wrote his delightful collection of stories, *The Canterbury Tales*, late in the fourteenth century—more than six hundred years ago. In all that time, generations of English speaking peoples have spoken to their parents and their children, to their grandparents and their grandchildren; an unbroken line of understanding. Here are the opening lines of Chaucer's book:

Whan that Aprille with his shoures sote
The droghte of Marche hath perced to the rote,
And bathed every veyne in swich licour,
Of which vertu engendred is the flour;
Whan Zephirus eek with his swete breeth
Inspired hath in every holt and heeth
The tendre croppes, and the yonge sonne
Hath in the Ram his halfe cours y-ronne . . .

While there is no doubt that this is English, it is certainly strange. Looking more closely, we can see that several words stand unchanged, while others are readily recognized despite differences in spelling, words such as "whan" or "Aprille." Still others are utterly different: "eek," for example, meant "also," and "sote" was "sweet." "Swete" also meant "sweet." I have no idea what nuances distinguished them, but the fact that we have lost "sote" suggests that life is but half as "swete" as it used to be.[20]

In any case, this passage points to profound changes in our language which would, nevertheless, have been so gradual as to have been indiscernible. Constancy is a false impression. Things may look the same, despite the fact they are changing all the time. Recall Lyell and Gibbon's tale of the seven sleepers.

We may also recall Cuvier's bold vision of the earth's history and the importance of the fossil record. His fine read of swirl and contour enabled him to recognize distinctive fossils that could tag a bed of rock wherever you met it. His mastery of comparative anatomy enabled him to reconstruct ancient life forms from the mere fragments fossils so often present and bring them to life to the extent that only catastrophic revolutions could explain why they no longer hunted in the plains or grazed in the prairies.

But when we turn to Darwin's *On the Origin of Species*, we wait and wait, but fossils—the only direct archive of life's past—never make an appearance. And though when we visit a natural history museum today we are likely to find fantastic fossil forms of ancient life as proof that life has evolved, Darwin does not make this argument. His discussions of them in *On the Origin of Species* are essentially to deny that they are sound evidence for understanding life's history.

This is intriguing. It's as if you had tried to write a history of ancient Greece without using any Greek artifacts or documents. And it takes us right to the heart of Darwin's work and the extraordinary shift in his approach. Darwin writes of the past by examining the way things change in the present. His take is Lyell-like. He does not evoke an ancient landscape. He does not seek the character of a long lost moment. But what would such a history look like? Well, it looks like short-legged sheep.

The English countryside, especially in the North—Yorkshire and Northumberland— is criss-crossed by dry stone walls. Gently rolling hills and craggy mounts are laced by low walls of rock held in place by gravity, not by mortar. It is a practice which rises out of time immemorial, and it solved a problem: a solution modern science verifies. Sheep, it seems, are victims to parasites with complex life cycles that include fairly long-term stages as cysts or larvae living in the soil. To ensure the well-being of the flock, a farmer needed to parcel the land and rotate its use, making sure that fields lay fallow long enough—for years, in fact—so that the parasite would die out before the land was used again. Pretty clever, but also very demanding when it came to making the stone walls which sectioned the land.

Hence the value of short-legged sheep. The walls don't have to be so high, because the sheep can't jump as high. The English, it seems, were rather proud of their breed of sheep with short legs, an eminently practical solution to a most practical problem. What bearing do they have on the concept of evolution? It is with short-legged sheep that Darwin begins his argument for evolution.

The fact of domestic breeding establishes the pliability of life. Whether it was sheep, horses, pigeons, or cattle, Darwin saw in domestic breeding a wedge between the clarity of "plant a carrot, get a carrot" and the play of generations. Here were clear instances where the profiles of populations had been modified.

The same process applied to plants, by the way. The American Indian transformed a weed with a rather substantial white root—we call it Queen Anne's Lace—by only allowing certain large rooted plants to go to seed. The result, after many generations, was a rather different looking plant with an orange root: the carrot. In other words, what Darwin realized was "plant a Queen Anne's Lace, get a carrot."

In this way, Darwin shifted the problem, moving the analysis of life's history from the fossil record to heredity and processes of change, those evolutionary bits happening now all around us. For Sedgwick the creation of new species that would succeed was beyond anything we could see in nature, and pointed to the finger of God. For Herschel it was the mystery of mysteries. But for Darwin, it was neither a miracle nor a mystery. It could be seen in something as prosaic as short-legged sheep. There is a pliability to life; a little of this, a little of that. Despite the intricacy of the coordination of its parts so brilliantly explored by Cuvier, there was some "slack," some "play," sufficient for careful husbandry to "make" short-legged sheep, draft horses as well as thoroughbreds, and even carrots.

Like Grote and Lyell, Darwin anchored his account of the distant past in a careful analysis of "present facts" and "causes now in operation." There is no reason to link his approach directly to Grote or the Benthamian circle. We have already noted Huxley's observation about the unity in method in Grote's *History* and Lyell's *Principles*. And we know Darwin carefully studied Lyell's *Principles* aboard the *Beagle*.

FIRST YEAR ON THE *BEAGLE*

This is the central value of examining scholarship in 1832 with an eye on the young Darwin. Thirlwall and Grote offer us a lens for viewing the cutting edge of historical analysis in this era. It served us well in our study of the powerful work of both Sedgwick and Lyell in geology. Now we come to Darwin. He had been delighted with his fieldwork with Sedgwick, and taken with the idea of becoming part of the inner circle of the Geological Society. But he also had Lyell's *Principles* with him and lots of time to weigh its argument and get the heft of Lyell's analytical "hammer." We might see this as having the best of both worlds, but at heart a choice had to be made, and it was made early on.

On January 6, 1832, the *Beagle* reached Teneriffe which was to be their first stop, but they were not allowed to disembark. The authorities feared they might have cholera on board (faint splash from that first bucket we drew from the English sea in chapter 1).[21] So Darwin had to wait another several days before he could leave ship in the Cape Verde Islands, about 350 miles off the coast of West Africa, and begin his geological studies. He focused on Quail Island, a small island off Santiago, or St. Jago as he called it: the largest island of this archipelago.

He saw a complicated tale in these two volcanic islands:

- they had originally been submerged beneath the sea, as evidenced by marine beds atop igneous rock;
- then an additional layer of lava had been added;
- followed by uplift, and a partial sinking.

Darwin's notes and sketches are competent and reflect his Cambridge schooling. They also reflect an appreciation of the light these islands cast on a matter of some dispute at the time, the character of uplift. And here, with his first opportunity as naturalist aboard the *Beagle*, we see Lyell's impact.[22]

Consider first the frontispiece of volume one of the *Principles* (figure 8.1), a drawing of the ruins of a Greek temple in the town of Pozzuoli in the south of Italy, on the Gulf of Naples. Lyell's interpretation of the bands marking the three ancient columns of the temple was a quintessential illustration of his method.

The bands are the product of burrowing sea life which typically lives in the intertidal zone. Hence the columns were initially above sea level when the temple was built, then below sea level at some point, and then again above sea level.

In 1826 Charles Daubeny, a professor at Oxford, had written an account of the volcanos of Europe. He discusses the columns at Pozzuoli and frames the matter nicely. While the bands could readily be explained by changes in sea level, other coasts of the Mediterranean do not show the same sorts of changes. However, if we seek to explain it via the lowering of the region some thirty feet and then the subsequent uplift, we run into another problem: "Had such been the case, it is probable that not a single pillar of the temple would now retain its erect posture to attest the reality of these convulsions."[23] With nowhere else to go, Daubeny opted for a salt water lake which had been created by volcanic activity and then subsequently gave way. An interesting hypothesis, but one Lyell denied.

For Lyell, Pozzuoli is another instance of Rye–Aldeburgh, instead of convulsions. That is, he sees the alteration of coastlines due to uplift and subsidence occurring at such modest rates that the columns have been left standing—exactly unlike a Hollywood disaster film. To underline this point, he cites several examples of marked changes in land level that left buildings undisturbed—in India, Jamaica, and Chile—and then closes his discussion with excerpts from contemporary accounts of extensive volcanic action in Pozzuoli in 1538 where the authors describe clear evidence of uplift without alluding to buildings collapsing.[24]

In his discussion of Santiago (St. Jago), Darwin echoes Lyell: "Dr. Daubeny when mentioning the present state of the temple of Serapis. doubts the possibility of a surface of country being raised without cracking buildings on it.—I feel sure at St Jago in some places a town might have been raised without injuring a house."[25] Darwin had made his choice.

By April of 1832, Darwin had conceived of a larger project. Again, it rested upon a quintessentially Lyellian argument. One of the most striking phenomena of the open

Present state of the Temple of Serapis at Puzzuoli.

FIGURE 8.1 Ruins of the Temple of Serapis by Canonico Andrea de Jorio, 1820, and used by Lyell as the frontispiece of vol. i of his *Principles*, 1830.

sea is the coral islet or atoll, circular expanses of sea cordoned off by coral reefs. Coral is a soft-bodied animal that builds the structures it lives in and upon. However, they can only survive near the surface of the sea, no more than a few meters below. Hence the reefs could not have built up over the ages from the sea floor hundreds and thousands of feet below. How then were these atolls constructed? The key was subsidence. Inspired by Lyell, Darwin argued that coral living near the surface could be balanced by regular and modest subsidence, yielding a steadily growing reef.

Suppose you begin with an island, a volcanic peak in the proverbial middle of the sea. At a certain distance out from the shore, coral begins to grow. If the island begins to subside, the steady growth of the reef would keep the living coral near the surface. In time, the center could disappear altogether, leaving the reef as an islet or atoll. In this way, coral reefs become a lovely demonstration of long-term change due to indiscernible processes.

Years later Darwin would write: "I am proud to remember that the first place, namely St. Jago, in the Cape Verde Archipelago, which I geologised, convinced me of the infinite superiority of Lyell's views over those advocated in any other work known to me."[26] If he had followed in Sedgwick's footsteps history would have been quick to mark the influence of the master and the agency of his influence, but choice is always there. However impressionable we might take the young Grote to have been, when he first met James Mill and the philosophical radicals, the views of the Benthamians resonated with his sense for life, the universe, and everything else. He had found his "home." The young Darwin was in like position, and even though he had been quite taken by his foray into the Welsh countryside with Sedgwick, he would choose another's hammer after he read the *Principles*. It was Lyell and not Sedgwick that resonated.

BACK TO THE *BEAGLE*

The *Beagle* would not reach the coast of Brazil, the first extended stop in South America, until February 29. Meanwhile, back in Great Britain, tensions had been rapidly increasing. You may recall from our earlier discussion that in March of 1831 the first reform bill had narrowly failed; the Whig government had dissolved, and there were new elections. This election increased support for reform and a new bill passed the House of Commons by a significant margin in September of 1831. The voice of the voters was clear. This second bill, however, was soundly defeated in the House of Lords. This time the government was not dissolved. Instead, they temporarily closed the parliamentary session, and with the new session in December a modestly modified third version was presented and passed in the House of Commons by a wide margin in March of 1832—about the time the *Beagle* made its way to the coast of Brazil. And so things came down once again to the House of Lords. On April 14, the bill passed, but not really; many of the "yea" votes were cast with the clear signal that this was only on the condition that it would be severely modified in committee. Lord Grey refused the new committee version and the Whigs withdrew. Wellington set about trying to form a government—and Francis Place had an idea about using gold to stop him.

As there was tension at home, so there was tension abroad. Darwin tells of a sharp exchange with Captain Fitz-Roy over slavery. While in Brazil, the captain "praised slavery, which I abominated, and told me that he had just visited a great slave-owner, who had called up many of his slaves and asked whether they were happy, and whether they wished to be free, and all answered 'No.'" Then with a wisdom greater than his years, Darwin asked "whether he thought that the answer of slaves in the presence of their master was worth anything?" This upset Fitz-Roy deeply, but the quarrel was patched up. "Fitz-Roy showed his usual magnanimity," Darwin added. But we see here Darwin's liberalism, well-earned from his grandfathers, Josiah Wedgwood and Erasmus Darwin, both of whom had actively sought an end to slavery.[27]

A few days later, on April 14, Darwin met with slavery first-hand. In Rio de Janeiro, Darwin had met an Englishman, Patrick Lennon, who was on his way to his estate some hundred miles inland and he joined him on this journey. This trek would offer Darwin his first real excursion into the rain forest and he was thrilled. His journal is filled with his delight at the richness of the flora and fauna. Here is a snippet from a letter to Henslow, professor of botany at Cambridge and the man who recommended him to Fitz-Roy: "Here [Rio], I first saw a tropical forest in all its sublime grandeur— nothing but the reality can give any idea how wonderful, how magnificent the scene is. . . . I never experienced such intense delight."[28]

Yet while at Lennon's estate he witnessed a quarrel between Lennon and his manager. Lennon, in a fury, proposed to take the women and children from the male slaves and sell them at auction in Rio. Darwin remarked in his journal: "I do not believe the inhumanity of separating thirty families that had lived together for many years, even occurred to the owner."[29]

Later Darwin wrote to his sister Caroline: "I was told before leaving England that after living in slave countries all my opinions would be altered; the only alteration I am aware of is forming a much higher estimate of the negro character."[30]

The *Beagle* spent months working its way down the eastern coast of South America. Captain Fitz-Roy worked meticulously and that might well mean going back to double-check earlier findings. At one point Darwin stayed on land to explore the region, rather than backtrack with the *Beagle*. At another the *Beagle* and Fitz-Roy had to work hard to avoid becoming embroiled in revolution in Buenos Aires. In various forays along the way Darwin and others collected fossils, as well as interesting plants and animals. Finally, in the middle of December they reached Tierra del Fuego, one of their core research sites.

At this point we find striking passages in Darwin's writings. His descriptions of the Fuegians are remarkably harsh. "I could not have believed how wide was the difference between a savage and civilized man," he writes, adding that their language "scarcely deserves to be called articulate."[31] And again of a group of Fuegians further to the West: "These poor wretches were stunted in their growth, their hideous faces bedaubed with white paint, their skins filthy and greasy, their hair entangled, their voices discordant, and their gestures violent. Viewing such men, one can hardly make oneself believe that they are fellow creatures, and inhabitants of the same world."[32]

Intriguingly, these sentiments were echoed in that same letter to his sister where he begins with a rather jaunty tone, speaking of being introduced to Tierra del Fuego by a master of ceremonies: a gale force wind. But then Darwin goes on to write: "I feel quite a disgust at the very sound of the voices of these miserable savages."[33] It's as if he started with a customary light tone, reflecting the warmth of family life, and then his dismay just pours out.

The intensity of these remarks is surprising. Darwin met with quite a number of indigenous peoples on this voyage, and nowhere else is he so harsh. Moreover, others offer more upbeat estimates. In his account of the voyages of the *Beagle*, Fitz-Roy reminds his reader that Caesar had found the Britons painted and clothed in skins like the Fuegians.[34] We also have a few other accounts of European encounters with Fuegians. William Parker Snow recounts his experiences stemming from a voyage in the mid-1850s. As they approached Tierra del Fuego everyone was afraid, but they were soon disarmed: "In a few moments we had them on the most friendly terms, laughing and mimicking whatever we did." He goes on to add: "Are these the wild Fuegians we have been made almost to dread?"[35] The women showed a degree of modesty and decorum and the men were honest in their trade. These sentiments have been echoed more than a century later by the contemporary anthropologist Anne Chapman, who has spent decades with several Fuegian peoples.

How should we understand it? Darwin had spent nearly a year with three Fuegians on board the *Beagle* by this time. He certainly did not find them hideous. Indeed, he was most impressed by how quickly they had adapted to English customs and language, observing of Jemmy Button: "It seems yet wonderful to me, when I think over all his many good qualities, that he should have been of the same race, and doubtless partaken of the same character, with the miserable, degraded savages whom we first met here."[36] What had struck Darwin were not racial differences but the extraordinary gap between humans in the wild and in civil society. The contrast between the civility of the Fuegians on board and the barbarity of the Fuegians in the wild was stunning— so stunning that it became a core problem for the young naturalist, both intellectual and visceral.

At one point or another we all find ourselves outclassed by life. It might be by the threat of violence; a battlefield, or simply walking down a city street at night. It can be the loss of a loved one, a spouse, a child. It can even be vicarious. A tale can take us to a moment where we realize that we would not be able to cope, that we could but walk away and hope to survive the consequences. Literature is filled with such moments. In *The Red Badge of Courage* (Stephen Crane, 1895), *Lord Jim* (Joseph Conrad, 1900), *Something Happened* (Joseph Heller, 1974), *Sophie's Choice* (William Styron, 1979), and many more we watch as young men, fathers, mothers, find themselves overwhelmed, and while they may able to walk away, it is not without a heavy burden.

There was a powerful instance of this toward the very end of Darwin's voyage. The *Beagle* was on its way home in the summer of 1836, but Captain Fitz-Roy wished to double check some measurements. They stopped one last time along the Brazilian coast. Here is Darwin's note in his journal: "On the 19th of August we finally left the shores of Brazil. I thank God, I shall never again visit a slave-country. To this day, if I hear a distant scream, it recalls with painful vividness my feelings, when passing a house near Pernambuco, I heard the most pitiable moans, and could not but suspect that some poor slave was being tortured, yet knew that I was as powerless as a child even to remonstrate."[37] Powerless as a child; nothing to do but walk away.

It is clear from the vehemence of his remarks that Darwin had been struck forcefully by his encounter with the Fuegians. He was not directly threatened by violence at the hand of Fuegians, nor did he lose a loved one, but still there was both threat and loss, and as with the event in Pernambuco, all he could do was walk away. How often are our lives marked by a singular event or happenstance, perhaps a caustic slur hurled across a disagreement, a ripple of a disturbance that pulls back the curtain on some cherished notion? Don't we carry the scars of chance encounters like that deaf woman (my mother in fact), who stopped meeting people's eyes as she walked down the street? How vulnerable would we be finding ourselves at twenty-three, halfway round the world from home, sailing through storm-laden seas, and entering a domain where the struggle for survival is so overwhelming? How much would it take? How long before some event would leave a lasting scar?

What are we at heart? It is a commonplace to see two basic alternatives. Either people are selfish and only prevented from being wicked by the teachings of religion, supplemented by the threat of social sanction—the policeman's truncheon; or they are innately good, and given the chance to go on about their business without fearing for their very survival they are other-directed, caring for those around them.

Darwin was a member of this second camp, a good west-country liberal. He supported parliamentary reform and opposed slavery. He had looked forward to meeting the noble savage in his native setting. He had read lots of travel literature, much of which was of an openly liberal stamp, as we have seen in Joseph Banks's account of the Tahitians. As the *Beagle* approached Tierra del Fuego, Darwin braced himself for both the harsh climate of this desolate region and the reputed harshness of its native inhabitants. Even so, he was shocked.

He was shocked and he couldn't put it down. We may remind ourselves of Goethe's impulse to find some event that would have inspired the depth of feeling in Byron's poem "Manfred." What had Byron endured that led him to write so? Granted, Grote saw Goethe's impulse as an instance of humankind's near irresistible urge to invent explanations, often out of whole cloth, but it is a hard impulse to resist. Let us try with Darwin.

Consider this entry in Darwin's journal for late December 1832, not long after Grote had become MP for London and Sedgwick had delivered the sermon for the Trinity College anniversary. It refers to the account given by Lord John Byron, grandfather of the poet, who had been a naval commander in the latter eighteenth century. His ship had been separated from the fleet during a storm and he was stranded for two years on an island in Tierra del Fuego: "Was a more horrid deed ever perpetrated, than that witnessed on the west coast by Byron who saw a wretched mother pick up her bleeding dying infant-boy, whom her husband had mercilessly dashed on the stones for dropping a basket of sea-eggs!"[38]

A father hands his son a basket of eggs. The son stumbles, the eggs fall to the ground and break. The father then picks up his son and swinging him, dashes his head against the rocks, leaving him dying on the shore. This deed, though Darwin saw it only in his imagination, resonated with his deepest feelings about the Fuegians. So it is that what Goethe had sought in the poet Byron, we can find in Darwin's read of the grandfather. This most horrid deed seemed to capture for Darwin the differences between barbarity and civility. He would come back to it again and again.

Suppose we ask ourselves about the father. What does he tell us about being human in this setting? Here is a society without institutions—no government or law, no organized religion. There is nothing "out there" to help shape sympathy for one's fellows. They lacked, as Sedgwick had just put it in his *Discourse* sermon back at Trinity College, "those moral and intellectual qualities which bind men and families together, and form the very sinews of national strength."[39] Or as Bentham might have put it: there is no disposition in the state of things that rewards or cultivates sympathy. Alone in the wild, they confronted their needs guided only by instincts. They lived in small groups, constantly moving about, constantly facing hunger and harsh conditions. The mother clearly felt remorse, but the father showed none.

Experience is but a prompt for the reality seen by the mind's eye. That's why clarity is gained on reflection. Lying on a bunk in narrow quarters, trying to sleep, Darwin mulls over the day's, the week's experiences:

Third week of December, that's mid-summer here, what a contrast with mid-summer back at home, I'd just finished at school, and worlds away . . . We'd read Shakespeare aloud, laughing at blundering Bottom.

This may be the bottom of the world, but there's no story here only pathos, no grace only suffering endured, and the cold and the wet, the relentless cold and wet.

Guarded, suspicious, no hint of the broad laugh of Jemmy Button or the gentle smile of Fuegia Basket, just the dour distrust of York Minster magnified a thousand-fold A Fuegian is a Fuegian, but when is he not? —A Fuegian on board is both a Fuegian and not a Fuegian. So much for Aristotle's excluded middle.

How far is it from Tahiti to Tierra del Fuego . . . maybe five thousand miles . . . but the real distance is much greater, I knew this place would not be Tahiti, but I never suspected . . . So maybe they stole the Endeavour's sextant, but here they ransacked the house we'd built, stole everything they could lift, trampled out the garden we'd planted and threatened poor Matthew's life.

This was not idle curiosity nor a fascination with a trinket from an exotic civilization, a keepsake, no it was a desperate seeking after something that would relieve the struggle, some edge for survival There's a line somewhere, first comes survival then comes art, no art here, no civility . . .

The distance is echoed even more deeply in their sense of self, there's that lovely passage from Joseph Banks on women's clothing—a natural and simple dignity—matched here by that poor soul suckling her infant in the incessant drizzle, the sleet melting against her skin, no shelter from the cold for her or her child.

Nature here is no mere backdrop or setting, it is the enemy, an enemy they're powerless against They have no technology to speak of, they gather rather than make . . . and they curl up at night protected from frigid weather laced by fierce storms in feeble shelters of sticks and twigs and leaves.

At least they have fire, a smoking, smoldering wet fire without which they would have perished long ago It is not just technology they lack, they lack community, social structures, what language they have is more like that of coarse birds, sufficient to give warning of danger or to announce a readiness to mate, but insufficient for stories of the sort which bind scattered souls into a society by giving meaning to life and place within the group.

Sadly but most assuredly this is a portrait of our distant past when we had our own humanity to invent as we scrambled over ledges of coal seeking warmth, were we not then as they are now, naked painted shivering hideous savages who no doubt have fleeting glimpses of the way things might otherwise be but lack the language, and even the occasion for sharing such visions.

Without community, without the common campfire, without the support that comes from tales by elders recounting origins and giving account of the purpose of present practices . . . without these they are left with little else than struggle marked by desperation, what else but desperation would lead a father, seeing his son drop a handful of eggs, to dash his brains against rocks leaving him dead on the shore.

Whatever Darwin's actual musings as he lay on his bed late at night trying to make sense of what he'd experienced, he consistently wrote of the Fuegians in the wild in this manner. In his diary, his published *Journal of Researches*, and in his letters; the line "naked painted shivering hideous savages" comes from a letter he wrote to the Rev. Kingsley in 1862. It is the reality he carried with him from 1832 for the rest of his life. And with it he would observe of such wretchedness and such savagery: "Such were our ancestors."[40]

And so ends the first year of Darwin's voyage.

COMING HOME

Captain Cook's *Endeavour* returned to England on July 12, 1771, after nearly three years at sea. Joseph Banks was twenty-eight. The *Beagle* returned to England in October of 1836, a few months shy of Darwin's twenty-eighth birthday. Both Banks and Darwin had been transformed by their voyages and both built successful careers on the basis of their experiences. Yet there were fundamental differences.

We may go back to Escott's observations on the profound shift in the values of the landed elite across the early Victorian era, following parliamentary reform. Leadership, politically and culturally, was no longer theirs by birthright, and they began to earn their place by more fully engaging in their affairs and careers. It was not enough to be gentlemen. They had now to prove themselves.

Though not a peer, Banks's father was a wealthy Lincolnshire squire, and 1771 was long before the passage of the Reform Bill. On his return Banks found he was a celebrity, so much so that it led to an audience with the King, a meeting which led to a lasting relationship and the founding of Kew Gardens. Banks also parlayed his interest in the sciences into a decades-long leadership of the Royal Society. Yet he did so as a gentleman in the old sense. He was above the fray, using his influence to foster developments in one direction or another. He never completed the anthropological memoir he might well have composed; it was not what one did. It would have been too earnest—beyond grace. One might love music and even perform before family and friends, but you would be a patron to a chamber group, not a member.

The Darwins, too, were comfortably well-off and an accomplished family. Charles was a gentleman of some leisure. While on the *Beagle*, he had a servant. The only job he ever had was being a naturalist on the *Beagle*, a job for which he was not paid. He was able to support his family with his share of the family's holdings, later augmenting these with his successful writings.

When Darwin had graduated from university he had little sense for his future. He had spent time at Edinburgh University unhappy at the prospect of becoming a physician. Changing to Cambridge, he settled into the notion of becoming a clergyman, one feels, because it wouldn't get in the way.

Those several weeks with Sedgwick in the summer before he set sail had offered an exciting glimpse into the actual practice of a sophisticated discipline, and they planted seeds in his imagination. Geology became something more than theories he could appreciate. It was something he could do. Sedgwick had clearly liked him. He wrote Darwin a letter that September, responding to the news that Darwin was joining the *Beagle*: "I cannot but be glad at your appointment & I truly hope it will be a source of happiness & honor to you." Regarding preparation for the voyage, he suggested Darwin: "Go to the Geological Society and introduce yourself to Mr Lonsdale as my friend & fellow traveler & he will counsel you." He then offered several suggestions, including: "Study the *Geological Socys. collection* as well as you can—& *pay* them *back* in specimens—I am to *propose you* when the meetings begin."[41]

Darwin had become a "fellow traveler" and he was pleased, as in this passage he added toward the close of a letter to Professor Henslow: "Tell Prof. Sedgwick he does not know how much I am indebted to him for the Welch expedition.—it has given me an interest in geology, which I would not give up for any consideration.—I do not think I ever spent a more delightful three weeks, than pounding the NW mountains."[42]

Sedgwick and the Rev. John Henslow, professor of botany and Darwin's mentor, were good people. Geology was stimulating, the inferences subtle and their reach exceptional. And, being a clergyman would certainly not get in the way. Here was a respectable career, a good place in society, and the opportunity to throw yourself into an exciting science. His hammer would be a key part of his kit aboard the *Beagle*, and there would be three volumes of geology to match the five on the flora, fauna, and fossil findings of the voyage.

Nevertheless, things were very different for Darwin on his return from his voyage in October of 1836. He was no celebrity. Various findings, fossils, and specimens had been sent ahead along the way and had excited attention from scientists, but there would be no audience with Queen Victoria. Yet Darwin was the very opposite of discouraged.

He engaged in a whirl of activity. In his private life he became engaged, married, and found a place to live in London. His obligations to his work on the voyage meant spending considerable time preparing his researches for publication, both the technical material—finding experts to evaluate his collections—and writing a more popular account of the voyage, his *Journal of Researches*. By 1844 he had prepared nine volumes from the voyage, along with a significant number of papers. He also had become a secretary of the Geological Society. It was all very exciting.

But there was something else about these two voyages, these two men. Something deeper, more profound. It has to do with innocence. Journeys and loss of innocence are a stock in trade for the storyteller. Both Banks and Darwin journeyed around the world. Both spent time in exotic places far from the routines of English village life. And both were transformed by their experiences, but in very different ways. Rather than innocence lost, we might speak of Banks's experience as innocence gained. He comes away from Tahiti inspired by the grace of a more natural way of being. Darwin, however, had been hit hard. Rather than innocence gained, we might speak of dismay gained. So where Banks came home and settled into an active life where thoughts of Tahiti could drift across like a pleasant spring breeze, Darwin was burdened by the weight of his dismay. Thoughts of Tierra del Fuego were no spring breeze, more a cold wind urging him to make sense of how such could be our ancestors.

TELLING DARWIN'S STORY AND THE *TITANIC*

On top of all this busy-ness, he had also begun the work that would shape his career, opening the first of several notebooks on the transmutation of species in July of 1837. Starting these notebooks is itself an intriguing story.

Over the decades a standard account of Darwin's work has emerged which begins with his instincts as a scientist. As naturalist aboard the *Beagle*, he was inundated by an unimaginable raft of new facts: fossil finds, new species, and more. Commenting simply on the beetles of a Brazilian rain forest, Darwin wrote: "It is sufficient to disturb the composure of an entomologist's mind, to look forward to the future dimensions of a complete catalogue."[43] Take as a further example the famous fauna of the Galapagos Islands. Despite the fact that these islands are not far apart and offer roughly the same climate, varieties of many animals carry distinctive adaptations to habitats on particular islands. This geographical speciation is bedrock to evolution. Over time an original stock of finch, tortoise, and lizard has become a set of varieties on their way to becoming an array of new species.

As a teaching tool, the story of the Galapagos has become clean and lean, and a key piece of Darwin's own realization that life had evolved. He visited the islands in September of 1835, stayed for about a month, collected the specimens, noted their geographical distribution, and everything fell into place.[44]

In the 1980s Frank Sulloway noted, as no doubt had many others, that for so crucial a piece of the argument, Darwin had not given it much space. The finches did not figure in either the first edition of the *Journal* or in *On the Origin of Species*. Sulloway wondered if perhaps the story had it wrong.

With some inspired sleuthing, he was able to establish that Darwin had not in fact noted the geographical distribution of the finches. So now the problem shifted. If Darwin, in the course of a swift survey of several islands and their many unusual life forms, did not record where each of these birds lived, where did the geographical hypothesis come from?

It turns out that it came from John Gould, sort of.

In January of 1837, not long after the *Beagle* had returned, Darwin gives the birds and mammals he had collected to the Zoological Society of London. It is a fellow naturalist, John Gould, who examines the birds. After one week, Gould reports that the birds from the Galapagos variously seen as blackbirds, gross bills, and finches were all rather peculiar finches that constituted its own group of twelve or thirteen varieties. Darwin and Gould meet again in March to discuss further findings. Gould points out that the three varieties of mockingbirds Darwin had collected were actually distinct species. Further, here geographical information had been included, and each was unique to its particular island.

There is an intriguing passage in a small notebook where Darwin looks back on things as the *Beagle* sails home. This would be in 1836. At one point he puts together the accounts he had heard about locals being able to identify several varieties of tortoise in terms of particular islands in the Galapagos with the three varieties of mockingbird he had linked to three different islands; perhaps there is something here. If this sort of thing were generally true of archipelagoes, "such facts would undermine the stability of species."[45] The bold print here, however, is that this speculative note lies fallow until Gould's expertise gives it life. It took Gould to make geographical speciation a worthy conjecture.

Sulloway goes on to observe that just a few months after this conversation, in July of 1837, Darwin opened his first notebook on transmutation.[46] For Sulloway, the inference is clear. Darwin had only come to the notion that life had evolved after realizing, prompted by Gould, that species were sufficiently pliable that a population could diversify through adaptations to particular conditions. That's why the transmutation notebooks were begun after the *Beagle* had landed and not before. Here were the facts that opened Darwin's eyes to evolution and set him on his way, even if it had little to do with the finches.

A long time ago, now, there was an episode of "Hitchcock Presents" about a lonely middle-aged man, perhaps an accountant. (They always seem to be accountants.) He is on a cruise. It's a holiday. On board he meets an equally lonely middle-aged woman. We see them make tentative overtures to connect, and by the end of the program they are a couple. As the camera pulls away we see him with his arm around her, and we see that the ship is the *Titanic*. A simple, but well-told tale. We could have been told it was the *Titanic* straight off, but that would have deprived the tale of its punch because the *Titanic* is a closed story. We all know what happens.

This is a titanic moment. Darwin, too, is a closed story. If another young man had met with Gould then we would have been curious about why he saw in these facts an argument for evolution when Gould hadn't. But with Darwin we know what happens next. All we need is the prompt that sets up "and so he opened his first notebook on evolution."

But Gould was obviously familiar with these same facts, and we have no reason to believe that he also opened a notebook on the transmutation of species. Indeed, in volume three of the zoological volumes (1841) and again in the revised edition of the *Journal* published in 1845, Darwin discusses the Galapagos birds and Gould's observations. They failed to excite any call for evolution. Yet for Darwin this had been a turning point. Sulloway sensed there was a crucial issue here, noting: "Indeed, as long as it is believed that Darwin's eyes were opened by an unbiased reading of the book of nature, the most interesting source of his conversion is effectively obscured. That source is none other than Darwin himself; for it was he, and not the evidence per se, that ultimately imposed the unorthodox interpretations that led him to embrace the theory of evolution."[47]

"He and not the evidence *per se*." An intriguing way to frame what was going on. The conversations with Gould were crucial. Darwin's "unorthodox interpretations" were clear to him; but no one else could *see* them. The sharp divide between Darwin and Gould seriously undermines the standard account. It was not the *fact* of geographical distribution or other related sets of observations which called for evolution. It was something *in* Darwin.

RETHINKING DARWIN'S STORY

There are good grounds for the standard account. After all, scholars have been examining Darwin's work for well over a century. Here, for example, is a passage from Jonathan Weiner's *The Beak of the Finch*: "These islands meant more to him [Darwin] than any other stop in his five-year voyage around the world. 'Origin of all my views,' he called them once—the origin of the *Origin of Species*."[48] But if Darwin is telling us how important the birds of the Galapagos were, finches or mockingbirds, why would we press for something else? Who would know better?

Thirlwall had claimed that he could know Greek history better than the Greeks themselves, because modern historical criticism was so much richer than that of the ancients, but that's not what's going on here. There is no such chasm between the Victorian era and our own. Rather, it is the basic notion that claims are rarely as straightforward as they might appear. This was the argument put forward by an intriguing scholar, R. G. Collingwood, some seventy-five years ago. For Collingwood, the act of understanding is directly tied to figuring out the question a person is thinking about as they write or say something. He came to this out of an interest in archaeology where he found himself routinely looking at artifacts and asking what problem they had been designed to solve. Only by knowing what it had been for could you understand what it meant, and with that, what life had been about for these ancient peoples.[49]

Collingwood proposed that we do this all of the time. Early on we develop a facility for reconstructing what is on someone's mind by trying to put ourselves in their shoes, as it were. We ask ourselves: what would I mean if I were to say such things?

This was a rather unfashionable approach to logic back in Collingwood's day, given the extraordinary success of mathematical logic and the work of Russell and Whitehead among others which looked at arguments and patterns of inference that preserved the truth of propositions, independent of what they were actually about. But it wasn't the truth of the proposition that concerned Collingwood. One could always provide a truthful description of some ancient artifact. It was its meaning: a shadow of John Stuart Mill on the differences between Bentham and Coleridge.

What does Collingwood's search for meaning look like? As it happens, I can offer an illustration straight out of my life. Many years ago now, I had a curious conversation with my son. He had just turned four and, as I tucked him into bed for the night, I said that he was old enough now so that pretty soon he would be able to start at the Park School, where I was a teacher. After a somewhat lengthy pause, he told me that he liked being four. Taken aback by this curious response, I said I was happy he enjoyed the prospect of being four and I was sure he would like being five and six and so on.

Something was up, but I would not figure it out until a month or so later when Pete spent a day at Park. We picked him up at the end of the day and he had this big grin on his face. "They have little kids!" he exclaimed. You see, I taught in the high school, and Pete must have assumed that when I said he could go to Park that I meant he would go to the high school. A little concerned about his father's judgment, he had offered the telling observation that he liked being four. Unfortunately, I hadn't heard it; I had no idea what he was worrying about and Pete had to suffer my lack of understanding.[50]

It wasn't a question of whether or not it was true. Pete very likely did look forward to being four; it was what it was really about. We are always archaeologists, à la Collingwood, trying to reconstruct intent and purpose from what is presented to us. So let us look at Darwin's observation this same way, seeking the question he would have been asking himself, so that "origin of all my views" was the answer.

For the standard account, the question would run something like: what evidence, what facts first alerted Darwin to evolution? But I think another question was worrying Darwin, a more derivative question that would have run something like this: supposing life has evolved, what processes might carry the promise of accumulated changes sufficient to separate plants or animals into distinct species from a given stock? The Galapagos Islands are thereby the origin of all his views on the mechanisms of differentiation. Or to go back to the language of John Herschel, the origin of all his views on the intermediate causes of the creation of new species.

From this perspective the Galapagos birds are a watershed for Darwin and not for others because for some time, years I believe, headwaters had been gathering, pressing for some account of evolution because . . . But we haven't uncovered, yet, what led him to such matters.

We know something is missing from the picture, something that separates Darwin from Gould. That Darwin is a closed story has masked a key issue. What was it? What made Darwin different?

HOW DOES YOUR GARDEN GROW?

Perhaps we can usefully borrow a notion from ecology. Some time ago a new island was raised on the Mid-Atlantic Ridge, just off the coast of Iceland, the island of Surtsey. Since 1967 its barren and craggy volcanic surface has given way to a variety of plants, over thirty species now, and a smaller number of animals. Initially only a few of the most hearty plants, like mosses, could take hold in the volcanic soil. As they flourished they trapped moisture, generated detritus, and all in all started the process of creating soils rich enough for less hearty plants. This has since been repeated to the point where there are now bushes and trees on the island.

The analogy suggests we shift our gaze to the suitability of soils, and with this a new "landscape" appears.[51]

The profound differences in what the facts *meant* to Gould and to Darwin point toward an active soil preparation. Darwin had a far richer read of the evidence. He was ready; he had worked the soil through, dwelling on the issues, mulling over the problems, developing prejudices and relishing predispositions. In this way, Gould's judgment that the three varieties of mockingbird were really distinct species could transform a "notion" into an hypothesis; the groundwork had already been done.

Evolution was not a new idea in 1832, or in 1837. Darwin's grandfather, Erasmus Darwin, had proposed a theory before Cuvier's work and Lamarck, a colleague of Cuvier's, had offered an important theory early in the century, as had Geoffroy St. Hilaire and Goethe. The idea was there, but it had not taken hold. Why not? The simple fact of the matter is that seeds work. Dogs continue to give birth to puppies and cats to kittens. But if seeds work, evolution can't get off the ground. There was a stalemate, with "plant a carrot, get a carrot" standing its ground against Spencer's "10 million acts of creation."

We know Darwin changed this. *On the Origin of Species* made evolution a most respectable hypothesis that would steadily move to the center of the biological sciences long before DNA and the mechanics of genetics would be worked out. But the question remains: what made Darwin receptive soil?

In a nice article, James Costa considers the leading philosophical issues of the day within biology and also points to Professor Henslow's sophisticated approach to botany, but these only carry so far.[52] After all, no one else saw what the animals of the Galapagos really meant. There was something else. What had Darwin been working at, such that sterile ground became fertile?

In fact, Darwin tells us.

ROSEBUD

A simple chronology should suffice. We have noted the swirl of work and writing upon his return to England: there was his *Journal* and eight further volumes on the flora, fauna, and geology of the voyage by 1844. Darwin continued his researches and reflections on the possibility that life had evolved and in 1859 *On the Origin of Species* was published. Darwin had not addressed the matter of human evolution in the book, but it was a matter long on his mind. By 1871 he was ready to take the proverbial bull by its horns and published *The Descent of Man*. Given the controversy that *On the Origin of Species* had provoked, Darwin knew this study would be provocative.

It is a lengthy text, and its last chapter recapitulates its major arguments. We come then to the last pages of this last chapter. The sigh is almost audible as Darwin opens the penultimate paragraph of his views on human evolution: "The main conclusion arrived at in this work, namely, that man is descended from some lowly organized form, will, I regret to think, be highly distasteful to many." "But," he then asserts, "there can hardly be a doubt that we are descended from barbarians." What follows takes him back almost forty years, as he conjures up his first encounters with natives in Tierra del Fuego: "The astonishment which I felt on first seeing a party of Fuegians on a wild and broken shore will never be forgotten by me, for the reflection at once rushed into my mind—such were our ancestors. These men were absolutely naked and bedaubed with paint, their long hair was tangled, their mouths frothed with excitement, and their expression was wild, startled and distrustful. They possessed hardly any arts, and like wild animals lived on what they could catch; they had no government, and were merciless to every one not of their own small tribe."

Here in this last reflection, Darwin puts what it means that we descended from apes in a context set by the meagerness, and cruelty of Fuegians in the wild. "He who has seen a savage in his native land will not feel much shame," Darwin wrote, "if forced to acknowledge that the blood of some more humble creature flows in his veins." He then refers back to two stories he had told in the body of his study: "For my own part I would as soon be descended from that heroic little monkey who braved his dreaded enemy in order to save the life of his keeper, or from that old baboon, who, descending from the mountains, carried away in triumph his young comrade from a crowd of astonished dogs . . ." Here is the nobility of the untutored primate soul. No threat of social sanction, no set of mores taught at church or school. There is something within primates which calls them to act in a manner directed toward the well-being of others. And now, Darwin closes the comparison by going back to that horrid deed: I would as soon be descended from monkey or baboon "as from a savage who delights to torture his enemies, offers up bloody sacrifices, practices infanticide without remorse, treats his wives like slaves, knows no decency, and is haunted by the grossest superstition."[53]

What an extraordinary passage. It is especially striking when we consider the concluding remarks in Darwin's many other works. Darwin closes *On the Origin of Species* with a "Recapitulation and Concluding Remarks," and he has concluding chapters in his many other works both before and after the *Origin*. In every other instance he summarizes his analysis, reviews relevant commentary of others, and reflects on interesting questions for further study. Only in the *Descent of Man* does he step out from behind the curtain and speak so directly, so forthrightly about what it all adds up to. There is no passive voice here, no effort to cast the matter impartially. It is his voice, his experience, direct and visceral.

We see here the push behind a whole lifetime of work. This is the problem which had demanded a solution. Darwin had believed that deep inside people are good. Virtue need not be imposed from without by threat of dire consequence. The human character is other-directed and virtue flows naturally—if we will but let it. Yet there was the savagery of those poor wretched Fuegians. The reflection at once rushed to his mind, "such were our ancestors." Are we all, innately, so coarse and cruel?

Reflecting for a moment on the link between Darwin, Grote, and Lyell, it seems somewhat irrelevant here. A line of analysis, a strategy for the conduct of thinking things through, is hardly the full measure of a person. In so many ways Darwin was different. Grote could hardly have been more alienated from his parents. Recall that his father was anxious to be a squire, his mother to be a pious recluse, and he referred to his home as clouded over by the "deepest night of ignorance." He, in turn, prayed to be free to exercise his own judgment—not to be free of onerous duties, but to choose the challenges he thought worthy. Lyell, too, grew up anxious to be his own man, free from the petty pushes and pulls that kept him from announcing his own path.

Then there is Darwin, whose childhood had been so rich and full its touch seems never to have left him. There was family everywhere—reading aloud, laughing, sharing passions and adventures, running, riding—a whirl in every direction. Where Grote and Lyell wished to be left alone, Darwin could not imagine why. He was an affable soul. Life was a good fit. Grote got up at 4 a.m. to study, because his father thought such pursuits idle. If Darwin awoke at 4 a.m., it would be to meet with a small cohort to see what it was like to climb a tree in the dark or measure the temperature of moonlight.

If Grote or Lyell had sailed on the *Beagle*, I dare say they would have seen the Fuegians as yet another chunk of human misery. It's not that they were indifferent. We know, for example, that Grote was passionate about the ills of society and had been devastated by the loss of his child and the near loss of his wife. But they knew indifference in others, and from a platform marked by such indifference how large a step is it to expect meagerness or cruelty in others? That we had such savage ancestors would not have been so surprising. For Darwin, however, Fuegian manners and mores were unimaginable and he couldn't put them down. They stayed with him for forty years.

"It is strange how fear resists the attacks of reason." This is a line from a memoir by John Updike. It's about Halloween and the way a child will be frightened by a mask even when he knows who's behind it.[54] It seems curiously relevant here. The mask was frightening, but what was more frightening was the possibility that it was no mask at all.

AND SO HE PUSHED

Why would anyone devote any significant stretch of time, let alone over twenty years trying to prove such an extraordinary hypothesis? It's not like Darwin had stumbled across some item that showed him life had evolved. This is the deeper mystery of mysteries: What could ever have compelled him to work and rework this terrain? And now we have an answer: a savagery Darwin could not put down.

Long before that December in Tierra del Fuego a Tahitian had been fascinated by the intricacy of a quadrant and put the *Endeavour*'s entire mission in jeopardy. Joseph Banks had acted swiftly, risking his life to track down the thief. It was an impulsive act, tinged with the arrogance of an accomplished young man. In Tierra del Fuego, Darwin met with a different sort of violation, one that challenged his deepest judgments on the character of humanity. He could but walk away, but that was not the end of things. Too much was at stake. And with the arrogance of an accomplished young man, Darwin pushed and pondered.

I do not know if Darwin favored chapter and verse. I rather suspect not. But there are two passages from *Genesis* that seem right here. The first concerns Abraham.[55] The Lord shares with Abraham that the wickedness of Sodom needs to be punished. Then

Abraham does a remarkable thing. "Suppose there are fifty righteous within the city," he asks. "Wilt thou then destroy the place and not spare it for the fifty righteous who are in it?" God agrees that he would spare the city for the sake of the fifty. And then in a still more remarkable manner, Abraham pushes: what if there are forty-five or thirty? Or twenty? Finally the Lord agrees: "For the sake of ten I will not destroy it."[56]

I cannot cite the second passage. It isn't there. For when the Lord tells Noah that he will destroy all of humankind and virtually all living things with them, Noah does not balk or barter. He simply accepts what Abraham would have challenged.

In this regard, Darwin was Abrahamic. He could not accept savagery at the heart of being human.

ESSENCE OR ACCIDENT?

From July of 1837 to October of 1838 Darwin went through a number of notebooks. By looking at these and jumping forward to compare his jottings to the text of the *Descent of Man*, or jumping back to the events of 1832, perhaps we can gain some further insight into how Darwin had worked to build the soil.

The earliest notebook after his meeting with John Gould is the Red notebook, which had already been started toward the end of the voyage in May of 1836. After the Red notebook, Darwin opens notebooks A on geology and B on natural history. B is followed by C, D, and E also on natural history. These draw from his readings and include a broad array of items. Here, for example, is an item from B, perhaps July or August of 1837, where Darwin writes: "Countries longest separated greatest differences—if separated for immense ages possibly two distinct type, but each having its representatives— as in Australia." He goes on: "This supposes that in course of ages, and therefore changes every animal has tendency to change.—This difficult to prove cats, etc., from Egypt."[57] Already, Darwin is speculating on the range of phenomena evolution might address beyond the islands of an archipelago, such as the distinctive mammals of Australia. The last note regarding cats, by the way, goes back to an argument of Cuvier's, after he had dissected the mummified remains of a cat brought back from Napoleon's conquest of Egypt. The remains were approximately three thousand years old and Cuvier found it identical to living cats, adding that if the cat had not changed in such a long span of time, why think any longer span would be any different.[58] We see in just this little note the point–counterpoint within the broad reach of natural history.

Mindful that Darwin would have also been working through his *Diary* and notes from the voyage in preparing his *Journal of Researches*, we are not surprised to also find entries referring to Fuegians in the wild. Here, for example, are some observations from notebook C written early in 1838 that he takes back to Fuegians in the wild. He begins: "Let man visit Ourang-outang in domestication, hear expressive whine, see its intelligence when spoken [to], as if it understood every word said—see its affection to those it knows,—see its passion & rage, sulkiness & very extreme of despair." Then, Darwin continues: "Let him look at savage, roasting his parent, naked, artless, not improving, yet improvable, and then let him dare to boast of his proud preeminence.—Not understanding language of Fuegian puts on par with monkeys."[59]

A comment or two on this passage. First, the tone of these observations is like that at the close of *The Descent of Man*; to wit, that phrase "let him dare to boast of his proud preeminence." Further, I read the last sentence as positing that Fuegians don't really have a language and so that puts them on a par with monkeys. Here, by the way, is Darwin's description of their language from his *Diary*, the entry for December 18, 1832: "Their language does not deserve to be called articulate: Capt. Cook says it is like a man clearing his throat; to which may be added another very hoarse man trying to shout & a third encouraging a horse with that peculiar noise which is made in one side of the mouth.—Imagine these sounds & a few gutterals mingled with them, & there will be as near an approximation to their language as any European may expect to obtain." Others attest to Fuegians having a significant language; Richard Lee Marks, for example, says their language had more than thirty thousand words, though there were some limitations to it.[60] What is important here is Darwin's perception, not the fact. He certainly does not respond to Fuegian language and culture the way Banks had responded to the Otaheites. Darwin had been hit hard by his encounters with Fuegians in the wild.

Finally, that striking phrase "roasting his parent" is a reference to Fuegian cannibalism. Though again there is some question of whether these accounts were accurate, Darwin tells us that Jemmy Button had spoken of how sometimes in the winter they would suffocate old women by holding them over the smoke of a fire and then eat them. This was then independently corroborated by the captain of a sealing ship who had been told the same thing by a Fuegian lad, who added that "women were good for nothing" and "man very hungry."[61]

The persistent threat of hunger lay behind more than cannibalism. In the *Descent*, Darwin will also discuss infanticide, suggesting that it "originated in savages recognizing the difficulty, rather the impossibility of supporting all the infants that are born."[62]

It is possible that cannibalism and infanticide had been freely chosen, but Darwin finds this hard to accept. Instead he invokes what we may call Bergeret's lemma, after the central character in a series of novels written late in the nineteenth century by Anatole France on contemporary French society, the corruption of its ruling classes and with it, the Dreyfus Affair. Though France portrays a wealth of characters, the only sympathetic one is Bergeret, a university professor. Professor Bergeret is a master of classical studies and so out of step with the mainstream of academia in his day. He is also a socialist, and so equally out of step with the politics of his day. He is, above all, however, an appealing soul, uncorrupted, with a regard for the miseries of the human condition, and a perspective which is both wise and tolerant.

Hence we come to a scene in the last novel of the set where France is secure in the sympathy we have for old Bergeret, and the door is opened a crack. Midst the corruption of the elite and the pettiness of so much else, France gives us a glimpse of the answer. Bergeret is walking through a park with his daughter, who asks him: "Don't you believe that men are naturally good and that it is society that makes them wicked?"

Bergeret answers "No," and he goes on to explain: "What I see is that they are emerging painfully and slowly from their primitive barbarism, . . . The time is yet far distant when they will be kind and gentle to one another." Nevertheless, this much he can say: "That men are least ferocious when they are least wretched."[63] And with this crisp line the conversation turns to the promise of technology and the progress of society, for Bergeret believes that if mankind is to be delivered, it will be through the machine and more particularly through electricity: the spark that flashed from the Leyden jar, the intangible seized by human hands, bottled and distributed over the wires that cover the earth. But we should not follow Bergeret any further. The key for us is the lemma—that men are least ferocious when they are least wretched.

And life was certainly wretched in Tierra del Fuego. The water that flows in the channels amidst these islands is only 2° Fahrenheit above freezing during the long dark winter, and no season is quite free from frost; even in summer it can fall to below freezing. In his journal, Darwin tells of a violent snow storm that struck Joseph Banks and a small party, killing two men. It was in the middle of January (the equivalent of our July).[64] It rains or snows three hundred days of the year, often as part of violent storms. The Fuegians lived in small, inbred family groups of ten or twenty that had to move on every few days, as they exhausted what food they could scavenge in one meager spot after another, restless and irritable, constantly facing not-enoughness. They lived on the edge. Those horrid deeds that so haunted Darwin tell us more about the conditions the Fuegians faced than the essential character of humanity. Their savagery was accident, not essence.

This had always been the issue. That last recapitulating passage tells us squarely that such were our ancestors. But with evolution we are also descended from primates who are capable of virtuous behavior, and that was the key.

THE METAPHYSICS OF MAN AND BEAST

In July of 1838 Darwin opens two notebooks. He opens D as a continuation on natural history from C, but he also opens M—perhaps for metaphysics—which picks up on this thread on mankind and is entirely given to notes and observations on our minds, and morality. We find here many items from Darwin's father, who was a physician and evidently a most sensitive and acute observer of the human condition. For instance, after telling several of his father's stories about people who late in life remember long-forgotten events and songs, Darwin adds: "Now if memory of a tune and words can thus lie dormant, during a whole life time, quite unconsciously of it, surely memory from one generation to another, also without consciousness, as instincts are, is not so very wonderful."[65] He comes back to instincts quite a bit later, beginning with the observation that babies know what a frown means very early in life. He then adds: "If so this is precisely analogous or identical, with bird knowing a cat, the first it sees it.—it is frightened without knowing why—the child dislikes the frown without knowing why."[66] You can feel Darwin pushing to tease out the nature of instincts and how they shape experience.[67]

But before we go any further, we should pause to note that in March of 1838 Darwin had become secretary of the Geological Society. William Whewell was the president, having followed Lyell in the president's chair. Whewell has come up several times in our study. He is a key figure in Sedgwick and Thirlwall's Trinity College ensemble. He was also a key figure in the Victorian scientific community. The son of a Lancashire carpenter, he entered Cambridge via a scholarship and never left. The Master of Trinity for a quarter century, he was master, as well, of an array of sciences: professor of mineralogy, recipient of the Royal Society's Copley Medal for work in mathematical physics, and sufficiently wise in matters geological to pen an influential review of Lyell's *Principles* and become the president of the society. And these were not his major scholarly accomplishments, for he is chiefly remembered for his masterly work in the history and philosophy of science: his *History*, first published in 1837, and his *Philosophy* in 1840.

On the evening of February 16, 1838 Whewell gave his first presidential address. He began by acknowledging the work of Richard Owen, who had just been awarded the Wollaston Medal for his work on fossils, most especially examining those Darwin had collected from South America. Whewell noted that three of these ancient beasts were of special interest, as they share many features with distinctive animals of that region in the present day: notably the Macrauchenia looks to be a giant version of the Llama, the Toxodon is a giant Capybeara, and the Megatherium looks like a giant Armadillo.[68]

After various other remarks, Whewell took up the fossil finds of a monkey-like animal, using it to broach a core matter. We all recognize, he offered, the "gradation in form between man and other animals," and so it need not startle us to find this reflected in the fossil record. But, he continues, it "is a slight and, as appears to me, unimportant feature, in looking at the great subject of man's origin." Then he offers his central thesis: "It would be most unphilosophical to attempt to trace back the history of man without taking into account the most remarkable facts in his nature: the facts of civilization, art, government, writing, speech—his traditions—his internal wants—his intellectual, moral, and religious constitution." There is nothing, however, in "the present state of things" that offers any hint at the origin of this fuller measure of man, and so Whewell suggests the geologist may close his own volume and "open one which has man's moral and religious nature for its subject."[69]

Darwin takes up Whewell's challenge by seeking an account of man's moral nature anchored in science and the present state of things.

With M, Darwin jumps right into the larger proposition that man's intellectual and moral constitution can be derived from the behaviors and instincts of his primate ancestors. Whewell has done Darwin the favor of laying out the proportions of the task before him, and so within weeks of that presidential address Darwin is conjecturing on a broad set of issues related to the nature of instincts, their relation to experience and heredity, the qualities of other-directedness, and a host of other features that will soon lead him to the construct of social animals and the fuller complexities of the struggle for survival.

Darwin writes in M: "He who understands baboon would do more towards metaphysics than Locke."[70] What metaphysics did Darwin have in mind?

On the Origin of Species and *The Descent of Man* gave a new relevance to the animal kingdom and most especially our closest kin, apes, monkeys, and other primates. But just as in the *Origin*, where it is striking that Darwin did not use the fossil record to bolster his argument, so in a similar manner it is striking how little emphasis Darwin places on anatomy and physiology in the *Descent*. Whewell's presidential address explains why. No matter how close the physical characters of ape and human, apes have no civilization, no government, or art. Though he acknowledges the work of others, notably Huxley, on the extended similarities in the physical character of higher ape and humankind, he is more concerned with the behavioral side of things, and most especially morality. His analysis is really about establishing the kinship of moral and social instincts in ape and human.[71]

The tack Darwin takes turns on the crucial concept of social animals. We find in M observations on a recent work by Harriet Martineau and the question of whether humans have an innate moral sense. On the one hand, she "argues with examples very justly there is no universal moral sense." But then, Darwin continues, she "allows some universal feelings of right & wrong (& therefore in *fact* only *limits* moral sense) which she seems to think are to make others happy & wrong to injure them without temptation.—This probably is natural, consequence of man, like deer &c, being social animal."[72] Note firstly how Martineau's comment is essentially Bentham's utilitarian principle. Rather than positing an innate moral sense, like Whewell, Sedgwick, and many others, she sees pleasure and pain as the bases for what is moral and just. Darwin doesn't question this, but links it directly to the fact that we are social animals. In effect he derives utility from the instincts that must obtain in order for social animals to survive.

Darwin is breaking new ground here. There is nothing of this sort in any of the earlier notebooks, his *Diary* from the voyage, nor in his soon to be published *Journal of Researches*. There are two subsequent items in M on this matter. In the first he teases out what is implied in his "utility" hypothesis, positing that actions necessary for well-being "are those which are good and consequently give pleasure."[73] Nature takes care of its own. An animal's pleasures guide it to do what is good for itself. This is not only a matter of physical needs, but also applies, for social animals, to interactions within the herd, hive—or out on the street or village green.

The other item sees Darwin feeling his way toward the origins of a moral code. He supposes there would be vague feelings arising out of the sense that there was something we ought to be doing. He talks of "being obscurely guided" in the face of strong sexual, parental, or social instincts, so we devise some rule like "Do unto to others" or "Love they neighbor" as a way to resolve our contradictory feelings.[74]

In the *Descent*, some thirty-plus years later, these conjectures are crystallized. Chapter 3 begins with these words: "I FULLY subscribe to the judgment of those writers who maintain that of all the differences between man and the lower animals, the moral sense or conscience is by far the most important." He then moves on to the question of the root of moral sensibilities, a question that "has been discussed by many writers of consummate ability"; but, Darwin continues, "as far as I know, no one has approached it exclusively from the side of natural history."[75] And this he proceeds to do, starting with the proposition that "the social instincts lead an animal to take pleasure in the society of its fellows, to feel a certain amount of sympathy with them, and to perform various services for them."[76] To this point he is more or less within the range of Aristotle's notion that we are social animals, and the city allows us to become most fully human. But natural history takes Darwin to greater depths. A bit later he says man has likely retained from distant ancestors "instinctive love and sympathy for his fellows" and with this, a tendency to be faithful to comrades.[77] In another passage he links this to survival, observing that "individuals which took the greatest pleasure in society would escape various dangers; whilst those that cared least for their comrades and lived solitary would perish in greater numbers."[78]

Darwin has pushed his way through the mud and found solid footing, something he can build upon: "Any animal whatever, endowed with well-marked social instincts, would inevitably acquire a moral sense or conscience, as soon as its intellectual powers had become as well developed, or nearly as well developed, as in man."[79] The heart of his analysis is to trace the development of moral sensibilities from the most funda-mental qualities of life for social animals. He offers, for example, that "the feeling of pleasure from society is probably an extension of the parental or filial affections." The foundations of morality rest upon the most fundamental experiences.

Then Darwin raises his head to contemplate the fullest measure of this social–moral construct. If we allow for the strengthening of these habits arising from social life, these feelings of love and sympathy for one's fellows, then morality's "ought" will free itself from any pleasure or pain of the moment. "He may then say, I am the judge of my own conduct, and in the words of Kant, I will not in my own person violate the dignity of humanity."[80] Though Darwin has not invoked an "archetype in the prescient mind of God," he was quite capable of fashioning a stirring vision of virtue, a virtue whose origin resides *in the present state of things*.

Darwin had been excited by Gould's observations on the speciation of the mock-ingbirds. The difference between his and Gould's reaction was crucial, opening the door for us to the work Darwin must have done to prepare his analytical "soil." Now we see an equally excited reaction. This time to Whewell's address. As several months earlier with Gould, Darwin opens a notebook and in short order lays out a series of observations and conjectures that leads to fundamental aspects of his subsequent theory. Again we are led to invoke a well-worked soil. Whewell was a prompt, tak-ing Darwin back to 1832 and the musings Fuegians in the wild had raised about the essential character of humankind. More than a prompt, Whewell gave Darwin a well-formed direction.

There is a further nuance connected to Darwin's metaphysics of man and beast. It rises in a paper by John van Wyhe and Peter Kjaergaard, where they examine Darwin's encounters, in 1838, with Jenny and Tom, the orangutans of the Regent's Park Zoo. Van Wyhe and Kjaergaard offer a view of Darwin's approach to human-kind's origins: "Darwin's troubling experience of witnessing the hunter-gatherer tribes of Yahgans in Tierra del Fuego may explain his view of orangutans as of such direct relevance to humans and his unusual lack of discomfort to relate apes and humans. The Yahgans had convinced him that the distance between highly civilized humans and the most degraded animal-like savages was narrow indeed. And there-fore the much exalted differences between humans and animals were greatly exag-gerated. It was thus only a small and painless step further to see great apes as human cousins."[81]

Our study suggests a different set of inferences. While we certainly agree that Darwin was troubled by his experiences in Tierra del Fuego, the difference between civilized humans and the most degraded savages was hardly narrow. Darwin had been stunned, but he did not see the Fuegians of the islands as more animal-like. Rather, he saw their meagerness and ignorance as "accident" and not "essence," a thesis proved by Jemmy Button, Fuegia Basket, and York Minster.[82]

We also agree that Darwin saw the differences between humans and animals as greatly exaggerated, yet there is nothing in his observations on Jenny and Tom, nor of that monkey and that old baboon that would suggest they were, at their best, rising to the savagery of the Fuegians. Darwin found Jenny as charming as she was petulant, and the story of that old baboon was a noble one. Morality is not a monotonic increasing function, progressing steadily along a scale of being. The broad primate essence we inherited, the instincts which inform our nonconscious minds, was capable of engendering other-directed–moral behaviors in orangutan and baboon. As such, it guarantees that Fuegian behavior was not a true window into the human character, but rather a window into how wretched the conditions were in that most inhospitable land. Tierra del Fuego was so wretched, it was hard—literally—to be human.

It's not that there was but a small step separating man and beast, but rather that it was one that natural history, that intermediate causes could manage. From that midsummer encounter in Tierra del Fuego back in 1832 on, it had always been about the essential character of humanity. It's not that the Fuegians were "animal-like"—we are all animal-like.

IN SUM

When we read Darwin's *Journal*, when we read Captain Fitz-Roy's *Narrative*, everything is more or less as it ought to have been, except for the vehemence of Darwin's reaction to the Fuegians in their native land. Others have noted this, but even as they ponder it for a moment or two, they go on to something else.[83] Yet this was the pivotal moment, or rather it reflects the pivotal problem. It marks when Charles Darwin started to become the man who would develop the theory of evolution.

Our thesis, our wooden horse, has five leading features. The first two frame the issue and the succeeding three constitute its explanation.

One: The standard account of the genesis of Darwin's work has emphasized the importance of the Galapagos Islands, an understanding bolstered by Darwin's remark that they were the origin of all his views. But the matter of geographical speciation only gained clarity when he met with Gould and the differences between Darwin and Gould highlight that Darwin was already receptive soil to the notion that life had evolved. This suggests that the importance of the Galapagos Islands had more to do with the processes of speciation than as a provocation of his thinking about evolution in the first place. It further suggests that the fact that Darwin has been a closed story has masked how much thought Darwin had given to these issues.

Two: Whewell's presidential address, like the differences between Darwin and Gould, points to a long-term thoughtfulness, such that within weeks of Whewell's argument that nothing in the present state of things could explain the origin of the core differences between man and beast—namely our reason and our moral sense—Darwin lays out the crucial construct of social animals and instincts for other-directedness.

Three: What, we may ask, had been the provocation that had spurred not a mere playing around with the subject, but a whole-hearted commitment over decades to tracking down vanishingly small wisps of change from one generation to the next? It must have been a most unsettling problem. In the *Descent*, Darwin tells us. He closes his lengthy examination of human evolution with a recapitulation of his argument, and he closes that recapitulation with a powerful, direct explanation of why he had for so long pursued the evolutionary hypothesis. Taking us back nearly forty years, he explains that he would sooner see humans as having descended from primates with their innate other-directedness, a built-in, ready-to-use, straight-out-of-the-box regard for their fellows that they share to varying extent with other social species, than to derive from savages who delight in torturing their enemies, offer up bloody sacrifices, practice infanticide without remorse, treat their wives as slaves, and are haunted by the grossest superstitions.

Four: That Darwin's reaction to Fuegians in the wild was the root provocation of his hunt for some framework that would assert the innate goodness (other-directedness) of humankind, leads us in turn to question why his reaction was so visceral. We have proposed here that his family life had been so vital and full of love that it shaped his sense that such was the natural state of things and that for it to be so profoundly otherwise suggests that life there had been cruelly harsh—that is, Bergeret's lemma.

Five: And finally, our examination of the brilliant historical scholarship in this early Victorian era, looking at Thirlwall, Grote, Sedgwick, and Lyell, suggests what analytical strategies were at the cutting edge as Darwin began to grapple with this profound and far-reaching set of issues, the complexities of Darwin's choice, and the contours of the analytical framework he relied upon as he set out to understand the true nature of the human character.

Events in Tierra del Fuego in December of 1832 imposed upon Darwin the fact that humans can be savage, but he resisted the inference that our nature was savage. He struggled against it and found an answer in evolution. To rescue humanity from a barbaric past, he graced it with a link to the rest of the animal kingdom. In the bravery and self-less virtue of monkey and baboon Darwin found a noble primate essence deeper than the outrages of a savage. In these deeds primates risked their lives for others. Here was the promise that humans are, at heart, good. We had inherited it from our fellow primates.

How ironic that so many have feared Darwin's work as a godless theory which debases our self-image. Darwin's aspiring ape was an earnest and virtuous beast. Benjamin Disraeli, novelist and prime minister, put it this way: "What is the question now placed before society with a glib assurance the most astounding? The question is this—Is man an ape or an angel? My Lord, I am on the side of the angels."[84] It turns out Darwin was, too.

Jemmy Button sensed that in Darwin; he brought him two spear-heads.

NOTES

1. Cannon, *Science in Culture*; Holmes, *Age of Wonder*; and Irvine, *Apes, Angels, & Victorians*.
2. Holmes, *Age of Wonder*, 26.
3. Banks, *Endeavour Journal*, entry for May 2, 1769, 77–78; see also Holmes, *Age of Wonder*, 5–6.
4. Banks, "On the Manners and Customs of the South Sea Islands," in *Endeavour Journal*, entry for August 1769, 124; cited in Holmes, *Age of Wonder*, 36.
5. Banks, "Thoughts on the Manners of the Otaheite," *Scientific Correspondence*, vol. ii, 331.
6. Franklin, "A Narrative of the Late Massacres, in Lancaster County," reprinted in *Gentleman's Magazine*, 34 (1764): 173–78.
7. The great French scientist Georges Cuvier concurred, seeing the voyage of the Endeavour as marking "an epoch in the history of science. Natural history contracted an alliance with astronomy and exploration." Cited in Holmes, *Age of Wonder*, 57 from Cuvier, "Eloge on Sir Joseph Banks"; translated and reprinted as "Historical Eloge of the Late Sir Joseph Banks," in *Edinburgh New Philosophical Journal*.
8. Darwin, "Letter to Caroline, March 30, 1833," in *Correspondence of Charles Darwin*, vol. i, 302.
9. Banks, *Endeavour Journal*, 106–07; cited by Holmes, *Age of Wonder*, 34. See also Banks, *Indian & Pacific Correspondence*, vol. i, xi.
10. Fitz-Roy, *Narrative*, vol. ii, 178.
11. See, for example, the Benjamin Subercaseux novel *Jemmy Button*.
12. Fitz-Roy, *Narrative*, vol. ii, 1–2.
13. *Ibid.*, 121.
14. *Ibid.*, 203.
15. *Ibid.*, 204, 210, 325.
16. *Ibid.*, 324.
17. *Ibid.*, 327.
18. *Ibid.*, 327.
19. Lyell, in Whewell papers, ms. #134.
20. Chaucer, *Canterbury Tales*, 81.
21. Darwin, *Journal of Researches*, 1. Hereafter cited as *Journal*.
22. See Herbert, "Charles Darwin as a Prospective Geological Author," for a richly detailed account of the importance of geology to the young Darwin and the character of his practice and the theoretical underpinnings of his work.
23. Daubeny, *Description of Active and Extinct Volcanos*, 162.
24. Lyell, *Principles*, vol. i, 452, 457–58.
25. Cited by Herbert, "Charles Darwin as a Prospective Geological Author," 170.
26. Darwin, "Autobiography," in *Life and Letters*, vol. i, 101.
27. *Ibid.*, 141.
28. Darwin, *Life and Letters*, vol. 1, 41.
29. Darwin, *Journal*, 23.
30. Darwin, "Letter to Caroline Darwin," May 22–July 14, 1833, in *Correspondence of Charles Darwin*, vol. i, 312–13.
31. Darwin, *Journal*, 195.
32. *Ibid.*, 203.
33. Darwin, "Letter to Caroline, March 30, 1833," in *Correspondence of Charles Darwin*, vol. i, 302.

34. Fitz-Roy, *Narrative*, vol. ii, 120–21.

35. Snow, *A Two Year's Cruise off Tierra del Fuego*, 324–26; see also Chapman, *European Encounters*.

36. Darwin, *Journal*, 2nd ed., 207–08.

37. Darwin, *Journal*, 499.

38. Darwin, *Journal*, 2nd ed., 246. The passage on Lord Byron's *Narrative* was not in the first edition. In the second edition the chapter on Tierra del Fuego is almost twice as long, with several passages that expand on the harshness of the land and its people. Byron wrote that a man "and his wife had gone off, at some distance from the shore, in their canoe, when she dived for sea-eggs; but not meeting with great success, they returned a good deal out of humour. A little boy of theirs, about three years old, whom they appeared to be doatingly fond of, watching for his father and mother's return, ran into the surf to meet them: the father handed a basket of sea-eggs to the child, which being too heavy for him to carry, he let it fall; upon which the father jumped out of the canoe, and catching the boy up in his arms, dashed him with the utmost violence against the stones. The poor little creature lay motionless and bleeding, and in that condition was taken up by the mother; but died soon after. She appeared inconsolable for some time; but the brute his father shewed little concern about it." Byron, *Narrative*, 148–49. This work was part of Fitz-Roy's library on board the *Beagle*.

39. Sedgwick, *Discourse*, 5.

40. Darwin letter to Rev. Kingsley, cited by Desmond and Moore, *Darwin's Sacred Cause*, 316. See also Darwin, *Darwin's Beagle Diary*, 120–44; Darwin, *Voyages of the Adventure and Beagle*, vol. iii, 227–45 (this is the first edition of Darwin's *Journal*); and Darwin, *Descent of Man*, vol. ii, 404.

41. Sedgwick, "Letter to Charles Darwin, Sept. 18, 1831," *Correspondence of Charles Darwin*, vol. i, 157–58; available online at www.darwinproject.ac.uk/letter/DCP-LETT-129.xml.

42. Darwin, "Letter to Henslow, May 18, 1832," *Correspondence*, vol. i, 171.

43. Darwin, *Journal*, 39. Regarding the standard account see, for example, the chapter on Darwin in Gillispie, *The Edge of Objectivity*, for a succinct and thoughtful account of his work.

44. See van Wyhe, "Where Do Darwin's Finches Come From?," 185–95; also Sulloway, "Darwin and His Finches," 1–53, and "Darwin's Conversion," 325–96.

45. Darwin, "Ornithological Notes," 264. The editor, Nora Barlow, dates this passage as September to October of 1835, while Darwin was in the Galapagos. More recent scholarship sees it as Darwin reflecting on his findings as he sailed home, June–July 1836. See also Barlow, "Letter to Nature, Charles Darwin and the Galapagos Islands," *Nature* 136 (September 7, 1935): 391.

46. Sulloway, "Darwin's Conversion," 389–90.

47. *Ibid.*

48. Weiner, *Beak of the Finch*, 4. The Darwin passage "origin of all my views" is in a personal diary or journal he began in 1838 and kept until December of 1881. It included back-filled information on significant events and efforts. This entry refers to an earlier period, July of 1837, when he opened his first transmutation notebook, now referred to as notebook B, and goes on to talk about the importance of the animals of the Galapagos. Van Wyhe, "Introduction to Charles Darwin's *Journal*," 13. See also Darwin, *Life and Letters*, vol. i, 142.

49. Collingwood, *Autobiography*, 29–39.

50. This is a story I shared in *Rethinking the Way We Teach Science*, 11–12.

51. This is called ecological succession; see Kormondy, *Concepts of Ecology*, 154–64. This concept, by the way, is traceable to Herbert Spencer. See van der Valk, "Origins and Development of Ecology," 37. This is less a new landscape than a renewed landscape; others have been here before. Take, for example, Francis Bacon's discussion of scientific knowledge and the need to clear our minds of various idols, which we can appreciate as an effort to characterize what we need to do to work our minds into suitable soil. See Bacon, *Novum Organum*, 19–35 of Book I. See also Karl Popper's discussion of Bacon and Descartes in "On the Sources of Knowledge and Ignorance."

52. Costa, "The Darwinian Revelation," 886–94; see also Costa, "New Landscapes and New Eyes," 42–55. See also *Darwin's Sacred Cause*, where Desmond and Moore also look for an underlying push behind Darwin's commitment to evolution.

53. Darwin, *Descent of Man*, vol. ii, 404–05.

54. Updike, *Five Boyhoods*, 173.

55. *Genesis*, chapter 18, verse 22.

56. *Bible: Oxford Annotated*, 21.

57. Darwin, "Notebook B," in *Charles Darwin's Notebooks*, 173 (corresponds to 15–16 of B online).

58. Rudwick, *The Meaning of Fossils*, 121–23.

59. Not long after opening notebook C, Darwin had his first direct encounter with an orang-utan. He spent much time at the Regents Park Zoo observing and interacting with their two orangutans. See van Wyhe and Kjaegaard, "Going the Whole Orang," *Studies*, 53–63.

60. Marks, *Three Men of the Beagle*, 6; *Penny Cyclopaedia*, vol. xi, 1.

61. Darwin, "Letter to Caroline," March 1833, *Correspondence of Charles Darwin*, vol. i, 303–04. See also Darwin, *Journal*, 2nd ed., 213–14.

62. Darwin, *Descent of Man*, vol. i, 134, and vol. ii, 363–65.

63. France, *Monsieur Bergeret in Paris*, 172–73.

64. Darwin, *Journal*, 2nd ed., 210.

65. Darwin, "Notebook M," *Charles Darwin's Notebooks, 1836–1844*, 2 (corresponds to p. 7 of M online).

66. *Ibid.*, 16–17 (corresponds to pp. 58–59 of M online).

67. For a charming set of reminiscences of his father see Darwin, "Autobiography," in *Life and Letters*, vol. i, 8–20.

68. Whewell, "Anniversary Address of 1838," *Proceedings*, vol. ii, 625.

69. *Ibid.*, 642. We may also add a line from Laura Snyder's excellent article on Whewell, where she quotes him from his work *The Plurality of Worlds*, written in 1855: "The introduction of reason and intelligence upon the earth is no part nor consequence of the series of animal forms." Snyder, "The Whole Box of Tools," 194.

70. Darwin, "Notebook M," *Charles Darwin's Notebooks, 1836–1844*, 23(corresponds to p. 84 of M online).

71. See Cannon, "The Whewell-Darwin Controversy," 380–82. Cannon suggests that much of Darwin's work in this formative period, the late 1830s and early 1840s, is a reply to the views of Whewell and Sedgwick.

72. Darwin, "Notebook M," *Charles Darwin's Notebooks, 1836–1844*, 21–22 (corresponds to pp. 75–77 of M online). The editor has added these two items from Martineau's *How to Observe*: "A person who takes for granted that there is a universal Moral Sense among men, as unchanging as he who bestowed it, cannot reasonably explain how it was that

those men were once esteemed the most virtuous who killed the most enemies in battle, while now it is considered far more noble to save life than to destroy it" (22). And: "Every man's feelings of right and wrong, instead of being born with him, grow up in him from the influences to which he is subjected" (23).

73. Darwin, "Notebook M," *Charles Darwin's Notebooks, 1836–1844*, 36 (corresponds to p. 132 of M online). Catherine Fuller shows that Bentham explored the notion of a sympathetic sanction to explain how people come to do good, in "Bentham, Mill, Grote," 125.

74. Darwin, "Notebook M," *Charles Darwin's Notebooks, 1836–1844*, 41–42 (corresponds to pp. 150–51 of M online).

75. Darwin, *Descent of Man*, 70–71.

76. *Ibid.*, 72.

77. *Ibid.*, 86, 137.

78. *Ibid.*, 80.

79. *Ibid.*, 72.

80. *Ibid.*, 86.

81. van Wyhe and Kjaegaard, "Going the Whole Orang," *Studies*, 51 (June 2015): 56–58. Desmond and Moore offer a similar view in *Darwin's Sacred Cause*: "But by lowering 'savage' morality and raising ape capabilities, Darwin made the continuum towards civilization seem more feasible" (370). But Darwin wasn't concerned about a continuum so much as what we are at heart—and so the importance of Bergeret's lemma.

82. See Darwin, *Descent of Man*, 34, where he makes the analogous observation in a passage about mental powers and the chasm between man and lower forms: "The Fuegians rank amongst the lowest barbarians; but I was continually struck with surprise how closely the three natives on board HMS *Beagle*, who had lived some years in England and could talk a little English, resembled us in disposition and in most of our mental faculties."

83. Marks, *Three Men of the Beagle*, and Chapman, *European Encounters*.

84. Benjamin Disraeli, Speech at Oxford, 1864, cited in Irvine, *Apes, Angels, & Victorians*, frontispiece.

9

AND SO . . .

A man walks down Piccadilly Street, playing over what had just happened at the Royal Academy and how he might tell the story of a splash of red paint.

A father in Leeds rises before dawn to make his way to the mill, worrying about his children and whether they will come down with cholera like the others in the neighborhood.

Francis Place stands alone in the garden, thinking about his friend, Jeremy Bentham, who has just passed away. *We're right in the thick of things now, but then we have been for a while. It might turn nasty; I'm sure the Old Duke is readying his men. Maybe it's good you've taken your leave. You were never one for pain or misery. But if it does work out, I'll be sorry you missed it. You would have loved turning that page.*

A man who built his chemical dye factory into one of the more successful industrial works in Manchester takes pride in his hard earned wealth, but worries about the unrest last summer in Bristol. The Reform Bill is coming up in the House of Lords again, and it's bound to be worse this time around. Perhaps he should pull his money out of the bank. They may have the right angle on things this time.

A popular young fellow at Cambridge with a gentle manner and a spark in his eye watches another issue of his journal make its way through the press, scanning the pages for errors. It's a modest run and he wonders how many more issues they'll be able to put out.

Meanwhile, an old school friend finds himself a candidate for Parliament. The whole thing is a bit of a blur, dashing off to meetings, comparing notes with Francis Place. At one point he pauses: *Lord knows there's much to be done, but if it took so long to get this modest reform bill through, what will it take to really change things? Being in Parliament might turn out to be more like spitting into the wind.*

Adam Sedgwick rises early this morning and makes his way along the Cam. *"Peaceful now,"* he muses. *"I've been here so long, the lecture halls, the rides across the fields . . . they carry too many names. How many lads have I taught? I should take a living somewhere. I could still do my work and find a wife, start a family . . . I'd miss the table talk, of course, but don't want to stay here forever . . . Ah well, best get on with things."*

And in the wilds of Tierra del Fuego, a young man works hard to distract himself. In some respects it's easy. He has collected so many specimens and they all have to be classified, labeled, and packed. Yet he finds it hard to focus. He's been at sea for about a year, but things are different now. He's not sure he can tell anyone about it, not even his sister. To describe it fully would be to bludgeon anyone, yet to tone it down would be too much like a lie. And he fears a greater sin. He might let it fade away.

In one of the short pieces collected in *The Garden of Epicurus*, Anatole France offers that philosophical systems are valuable for the study of man and the different conditions the human mind has passed through, adding: "They are like those thin threads of platinum that are inserted in astronomical telescopes to divide the field into equal parts. These filaments are useful for accurate observation of the heavenly bodies, but they are not part of the heavens."[1]

What about the lines of analysis in our study? Our telescope has been trained on three fields of vision: 1832, historical criticism, and Darwin.

1832

A year is an artificial construct. Days come and go. They add up. We shed their memory. On any day, somewhere it rains, somewhere a youngster catches her stride; a young man pauses to reflect on his life, his disappointments. In this everyday sense, 1832 was no different. The city could still lay claim to providing the paved path, however muddy or broken, that leads to the factory, to the church, the pub, and the cemetery. The countryside could still lay claim to the endless toil that turns fields into grain, cabbages, and their ilk. Life, no doubt, was as laden with burdens and hardship, as laced with possibilities and joy. Yet there was a difference.

Change pressed itself upon the day, and while the wisdom of the Preacher of Ecclesiastes tells us the sun rises and the sun sets, but the earth remains forever—nevertheless, the days of 1832 were charged with demonstrations in the streets, fires in the countryside, and drama in Parliament.

The hidden lemma in all this, well not really hidden, just more tacit than spoken, was industrialization. There's a cholera outbreak in Leeds and Manchester. The Industrial Revolution had displaced thousands of families from farm to factory, from tithed cottage to back-to-back, and when we look more closely at the epidemic we find that's just where it hit. Back-to-backs were an efficient use of space and materials, and just as efficient for the spread of disease.

The rapid rise of urban populations in Leeds, Liverpool, Manchester, Birmingham, London, and elsewhere across the country not only fueled epidemics, it also fueled stunning disproportions in representation in Parliament. There had long been pocket boroughs, but now a truly significant portion of the population lived in communities with little or no representation. This not only disenfranchised the workers; it also meant factory owners and merchants were without parliamentary voice. Hence the effectiveness of the run on the bank. Even Darwin's sojourn in Tierra del Fuego and the voyage of the *Beagle* related to industrialization, in that such voyages were part of England's work to secure imports from her far flung empire by securing safe routes and facilities where ships might make repairs.

Far more than the workplace had changed, urged Thomas Carlyle in a striking piece in the *Edinburgh Review*. The first flush of the Industrial Revolution had faded. What had been revolutionary had now become the norm, and Carlyle takes its measure. He begins his "Signs of the Times" this way: "Were we required to characterise this age of ours by any single epithet, we should be tempted to call it, not an Heroical, Devotional, Philosophical, or Moral Age, but, above all others, the Mechanical Age. It is the Age of Machinery, in every outward and inward sense of that word; the age which, with its whole undivided might, forwards, teaches and practises the great art of adapting means to ends."[2]

And so he sets out to trace these many outward and inward senses, as when he observes: "Our old modes of exertion are all discredited, and thrown aside. On every hand, the living artisan is driven from his workshop, to make room for a speedier, inanimate one. The shuttle drops from the fingers of the weaver, and falls into iron fingers that ply it faster. . . . Even the horse is stripped of his harness, and finds a fleet fire-horse invoked in his stead." This has not all been bad, however. "What wonderful accessions have thus been made, and are still making, to the physical power of mankind; how much better fed, clothed, lodged and, in all outward respects, accommodated men now are, or might be, by a given quantity of labour, is a grateful reflection which forces itself on every one."[3] And yet . . .

Carlyle moves on to the inward changes caused by the mechanical genius: "Here too nothing follows its spontaneous course, nothing is left to be accomplished by old natural methods. Everything has its cunningly devised implements, its pre-established apparatus; it is not done by hand, but by machinery. Thus we have machines for Education: Lancastrian machines; Hamiltonian machines; monitors, maps and emblems. Instruction, that mysterious communing of Wisdom with Ignorance, is no longer an indefinable tentative process, requiring a study of individual aptitudes and a perpetual variation of means and methods, to attain the same end; but a secure, universal, straightforward business, to be conducted in the gross, by proper mechanism, with such intellect as comes to hand."[4]

Carlyle sees the same organizational approach across the board. The individual has been swamped by mechanism in philosophy, science, art, and literature, and he adds that "these things, which we state lightly enough here, are yet of deep import, and indicate a mighty change in our whole manner of existence. For the same habit regulates not our modes of action alone, but our modes of thought and feeling. Men are grown mechanical in head and in heart, as well as in hand."[5] Carlyle sees men as having lost their belief in the invisible: "This is not a Religious age. Only the material, the immediately practical, not the divine and spiritual, is important to us."[6]

Carlyle's piece is stunning, with telling insight. England was becoming a mass society. It had always had its poor, but now they were quartered in such numbers and concentrations that traditional structures no longer supported them. The charity of the local lord of the manor no longer reached them and the ministry of the parish church was overwhelmed. New organizational strategies were tried: work houses, public health programs, even a profound recasting of the law. Such efforts acted on a different scale and for sure they would have been more impersonal, more mechanical, than the way things had been before.

Though Francis Place regretted John Stuart Mill's German metaphysics and his sympathy for the likes of Coleridge and Carlyle, yet his work affirms Carlyle's observations.

There had long been town meetings and local leaders who would petition the lord of the manor. As the workplace changed with the rising scale of the economy, combination laws and a common-law judiciary worked against any structural change in the voice of the workers. Place's push to undo these laws and allow workers to combine was, from his perspective, a natural adjustment restoring the worker's voice that had been drowned out by the din of mechanical shuttles, conveyor belts, and charging steam. In the push for reform, town meetings became marches with a hundred thousand strong, mass gatherings on an unprecedented scale.

Place would have appreciated this core to Carlyle's piece. But what it really does for us is to mark how the changes brought on by industrialization were now the object of thoughtful reflection. The immediacy of the revolution had now passed, giving way to reasoned evaluation: both praise and criticism.

We have borrowed from Anatole France both Bergeret's lemma and a telescope's platinum wires. Let us now borrow from another French savant, Gaston Bachelard, who offers an intriguing swirl of notions about images and the deeper workings of the mind. One passage in particular bears on our study. He writes: "Memories are motionless, and the more securely they are fixed in space, the sounder they are."[7] For Bachelard, the mind does not do anywhere near as well reliving duration as reliving setting or context. Memories are motionless; there's a solitude to them, and often they are heartwarming. On the other hand, action is threatening. This suggests that with a typical story, as with life in the wild, we fix on moving figures out of our anxiety. If we take away the motion, as with a posed photo or painting, things change. There may still be a central figure but without motion we can relax, and as we take in the image, the mind begins to fashion an interaction between the figure and its setting. We enter that space, forging an empathetic contemplation of our own.

The central historians in our study—Thirlwall, Grote, Sedgwick, Lyell, and Darwin—lived long lives fully engaged in the many issues of their day—from mandatory chapel to slavery, from expanding the Tripos to the secret ballot. We have acknowledged much of this work, but we have sacrificed the sweep of individual lives for a common narrow slice, that collective moment that was 1832. Bachelard has caught a key consequence here: it moves the focus to the "centers of fate," as he puts it; what we have called the internal push of their work, their nature as scholars. History, which sets aside the temporal beat of events and holds things still so as to dwell on setting, is able to evoke the contact mind makes with the press of a problem.

A dust mote's dance catches our eye—a random walk mapped by Bose-Einstein statistics. It points to a myriad of unseen collisions as the protagonist or particle interacts with its surrounding medium. What makes a Thirlwall or Grote, Sedgwick, Lyell, or Darwin more than a mote? Their scholarly lives are likewise filled with a myriad of unseen collisions between their sense for things and the surrounding medium, be it myth or fossil, colonial expedition of past or present, the critical sensibilities of an ancient commentator or the unsettling sight of naked Fuegians on a beach. The difference is their will, their agency—that decision, made in a flash, to give chase after a stolen sextant. But for our historians the chase is metaphor for pursuits guided by canons of criticism, and that is the second field we held before our telescope.

HISTORICAL CRITICISM

Our study moved from Reform to historical scholarship, to both classical studies and geology. Fossil and myth, what do they really mean? What do they point to? And what can they tell us about our past and the forces that have shaped our world, and who and what we are?

There emerged in the latter decades of what T. H. White called the age of scandal a broad historical framework, forged from the work of such scholars as Cuvier, Mitford, and William Jones. Sacred, civil, and natural history combined to tell a coherent tale from the geologically distant past through to the last great Cuvierian revolution, Noah's flood. From there, migrations lead to the resettlement of ancient lands, the subsequent teasing out of new languages and new national traditions, and on to Classical antiquity (recall that Sicyon was founded 259 years after the flood), which sets the stage for the Middle Ages and modern times.

As Parliamentary Reform sought to fundamentally reshape the politics of the old order, so did Thirlwall, Grote, Sedgwick, and Lyell seek to reshape its approach to history. And again, 1832 stands at the platinum wires, dividing the field into new quadrants. Thirlwall has reconceptualized ancient legends into myths, setting the table for a history of the ancient imagination instead of colonies, inventions, and military interventions; while Sedgwick has begun to apply the Eastern Alps thesis to the strata of Wales, which will lead to the concept of a geological system. Meanwhile, Grote has staked out his dramatic shift away from the history *within* Greek legend to the history of later commentaries and the emergence of critical thought, leading not only to Socrates, Plato, and Aristotle, but also to a democratic society; as Athens gains its Golden Age, Hesiod's vision of a fictional past becomes the way we see an historical epoch. And then there's Lyell's emphasis on a most prosaic present, turning revolution and discontinuity into a featureless replay of everyday occurrences.

While we may agree with Carlyle that this was a mechanical age and that industrialization had caused much disruption, our study has not witnessed the rise of a mechanical world view that dominated the modes of thought and feeling. Straight off, the work of Coleridge, Thirlwall, and Sedgwick was clearly spiritually minded and organic rather than mechanical. Nor should we set it aside as some sort of trailing residue of former times and sensibilities. John Stuart Mill assures us of that by seeing these Germano-Coleridgeans as reformers with worthy insights on how to move forward. Coleridge's clerisy was not the Church of England's clergy. Thirlwall's myth was not Mitford's legend. And Sedgwick's Eastern Alps thesis was not Cuvier's stratigraphy. History as driven by an underlying unfolding of ideas, reworked in the light of new conditions, applied forcefully in Sedgwick's geology, Thirlwall's antiquity, and Coleridge's *Church and State*.

The work of Bentham, Grote, and to some extent Lyell is more complicated. On the whole they were a skeptical lot, and if we borrow Disraeli's divide, then they were not likely to have been on the side of angels in any theological sense. Yet this doesn't mean they were without spiritual qualities.

Take George Grote. In his *Autobiography*, Darwin tells us what Carlyle thought of Grote's *History*, and we are not surprised. He thought it "a fetid quagmire, with nothing spiritual about it."[8] But we need to be careful. While Grote's *History* is not spiritual or religious in a traditional sense, it is a deeply humane work. His central thesis on the accomplishments of the ancient Greeks was not about economics or other material advances; recall Carlyle's line about how only the material and immediately practical was important in this new mechanical sensibility. The Greeks had not forged new weaponry to conquer their neighbors, but conquer them they did—by virtue of the draw of philosophy, of history, and the sciences. They had forged critical thinking from the pursuit of the deepest meaning of the "sacred" texts of their day, a pursuit that for Grote was hardly cynical or without passion.

In his *On Heroes, Hero-Worship, and the Heroic in History*, Carlyle discusses what religion really amounts to, writing: "By religion I do not mean here the church-creed which he professes, the articles of faith which he will sign and, in words or otherwise, assert; not this wholly, in many cases not this at all. . . . But the thing a man does practically believe (and this is often enough *without* asserting it even to himself, much less to others); the thing a man does practically lay to heart, and know for certain, concerning his vital relations to this mysterious Universe, and his duty and destiny there, that is in all cases the primary thing for him, and creatively determines all the rest."[9] In this sense, I do not hesitate to speak of Grote's *History* as the work of a religious man, a man who in his worldly affairs and in his scholarship was guided by quality not quantity, by moral–civic precept not profit, and by a regard for the well-being of all, not just the few.

DARWIN

Then there is Darwin. Modern science offers a mechanical view of life, fitting Carlyle's conception. Played out on a molecular scale—with its host of interlocking parts and the assembly-line action of tab A into slot B, as amino acids link to form proteins guided by templates registered in nucleotides. Evolution is mechanism par excellence. But that is not Darwin's evolution. The science of heredity in his day was all rules of thumb. Good enough to have produced a broad variety of domesticated plants and animals over thousands of years, but nowhere near the gear-works of modern molecular mechanics.

Darwin was a naturalist, not a biochemist, and his mechanisms played out in field and forest, not in test tubes or x-ray crystallography. The central dynamic was the biological equivalent of the notion traceable to Aristotle's *Physics*: "Nature abhors a vacuum."[10] Species will move toward "empty spaces," with adaptations that enable them to survive. This is the essential lesson of the Galapagos. Further, Darwin underlines that it is a blind motion. He sees heredity as fundamentally approximate. It is always shy of a perfect copy, more like the proverbial monk transcribing a manuscript than a Xerox machine. He postulates that there are bits of difference from generation to generation, and that they are heritable. If they prove to be advantageous, the individual is more likely to survive, and so it goes.

These proximate processes of heredity are the intermediate causes of evolution, but as we have seen with the differences between Darwin and Gould, such notions came after the fact. It was because he had been working and reworking the soil surrounding Tierra del Fuego that Darwin could see the promise of inherent genetic malleability in the curious creatures of the Galapagos.

In Tierra del Fuego Darwin had been stunned by the difference between savage and civilized man. Going back to Bachelard, this is our still life. Time has stopped; all we have is the steady contemplation of a problem. The image challenges us to enter that space and feel our way to Darwin's "center of fate"—his analytical "nature," that seedling growing out from his learning and his sense for how things are and ought to be, pushing itself forward, led by the press of a deeply felt problem.

If humankind is, at heart, a brute, then Tierra del Fuego is no problem. But it *was* a problem. We know because he tells us in his letters and later in his *Journal of Researches*, and much later he tells us again in that extraordinary passage closing the *Descent of Man*.

He also tells us in his construct of social organisms, which we find in notebook M started shortly after Whewell's presidential address. Harriet Martineau had claimed that there are nearly universal feelings about right and wrong that correspond to Bentham's principle of utility: it is right to make others happy, wrong to hurt them. In 1838 Darwin writes this is probably a natural consequence of man being a social animal—an instinct that must be there in order for humans to survive. Twenty years later in *On the Origin of Species*, Darwin would examine the structures of social insects, notably bees and ants, conjecturing as he had in his notebooks that natural selection would act upon instincts as well as other mental and bodily traits, as they are intimately connected to survival.[11] And again, in the *Descent*, Darwin will go back to the argument on instincts, seeing primates as social animals and rooting other-directedness to the core of their being.

As it happened, Darwin sent his beloved teacher a copy of *On the Origin of Species*, and Sedgwick had replied kindly and in earnest, reminding Darwin of their friendship but also fixing on what he saw as a fatal flaw. Darwin had abandoned final causes, trying to replace them with the struggle for survival, but for Sedgwick this would not do. It was God's purpose that linked matter and morality.[12]

Note that Sedgwick did not criticize Darwin for violating the letter or spirit of the Revealed Word, in keeping with his fundamental view that the Bible is a moral document, not a science book. It would be as mistaken, he suggested in his presidential addresses and in his *Discourse*, to look for science in the Bible as it would be to seek moral guidance from the laws of chemical combinations. But that, in effect, was what Darwin had done.

It was the heart of the matter for them both.

Thousands of miles from the cholera-infested back-to-backs of Leeds and the myriad of conveyor belts in its factories, a young man found himself on the shores of Tierra del Fuego with a shattered sense for the human condition. Nature had spoken to him as Yahweh had spoken to Abraham of a corrupt humanity, and like Abraham, the young man pushed back. There was good and it must somehow be secured.

Many another young man would have accepted nature's judgment. Certainly human-kind can disappoint, showing itself to be cruel and wanton. Such a man might have shrugged his shoulders and stowed away his disappointment. Another age, one less open than 1832, one more closed in its vision, might have been unable to nurture an Abrahamic response. How much can someone alone carry? But our young man had seen hard work and optimism carry the cause for Reform despite the vested interests of the Old Guard.

There was change and challenge as well in the world of scholarship, notably classi-cal studies, where Thirlwall and Grote were forging new understandings of the aim of history and new visions of what the ancients had to offer our sense of who we are. And Darwin had seen for himself the same promises in geology. He listened to Sedgwick as they traversed valleys and hills in Wales describing the beginnings of a new geology, one with a horizontal gaze that evoked the character of ancient landscapes and the plants and animals they had supported. But he would soon read Lyell carefully, absorb-ing the principles of his new geology so that they became his own.

In the end, it took all this. It took sensibilities that were vulnerable and open to the world, so that he could be deeply shocked. It also took the self-regard, the confidence to attend to his own convictions. Among these convictions was a thoroughgoing take on the elements of historical criticism. As the ancient Greek had seen the mighty arm of Hercules as the agency of history, clearing fields of beasts and boulders, so did Darwin see in Lyell's "causes now in operation" agents that had transformed the earth over eons.

Such were the components of Darwin's center of fate, which would warrant his decades-long search for proof that indiscernible change was both real and sufficient. Almost forty years after he stood on that Fuegian shore he laid out his account of the evolution of humankind.

A few grains of sand and a new vista was created as a massive arc of rock fell to the ground.

NOTES

1. France, *The Garden of Epicurus*, 117.
2. Carlyle, "Signs of the Times," in *Critical and Miscellaneous Essays*, vol. ii, 146–47.
3. *Ibid.*, 147.
4. *Ibid.*, 148.
5. *Ibid.*, 150.
6. *Ibid.*, 162.
7. Bachelard, *Poetics*, 31.
8. Darwin, *Life and Letters*, vol. i, 77.
9. Carlyle, *On Heroes*, 2.
10. Aristotle, *Physics, Works*, vol. ii, bk iv, 6–9.
11. Darwin, *On the Origin of Species*, chapter 8, "Instinct."
12. Sedgwick, "Letter to Darwin, November 24, 1859," in Clark and Hughes, *Life and Letters of the Reverend Adam Sedgwick*, vol. ii, 356–59.

BIBLIOGRAPHY

Archive Materials

British Museum, Additional Manuscript collection: unpublished manuscripts and notebooks of George Grote, miscellaneous correspondence of Babbage, Grote, Lyell, Murchison, Thirlwall, and Whewell.

Cambridge University: i. Adam Sedgwick Geological Museum, the Sedgwick papers; ii. Trinity College Library, the Whewell papers; and iii. University Library, Manuscript Room: Sedgwick papers, including unpublished autobiographical fragment, miscellaneous correspondence, and Grote papers.

Edinburgh University, University Library, Manuscript Collection: Lyell papers, including lecture notes for tenure at King's college.

National Library of Scotland in Edinburgh: miscellaneous correspondence of Lyell, Whewell, and others.

University of London, Senate House Library, Manuscript Room: Grote papers and volumes from his personal library, including notebooks and correspondence.

Printed Sources

Abrams, M. H. *The Mirror and the Lamp: Romantic Theory and the Critical Tradition.* New York: Norton, 1953.

Agassi, Joseph. *Faraday as a Natural Philosopher.* Chicago: University of Chicago Press, 1972.

Allen, Grant. "Sir Charles Lyell." *Popular Science Monthly* 20 (March 1882): 591–609.

Anon. "A Reply to Mr. Brougham's 'Practical Observations upon the Education of the People, addressed to the Working Classes and their Employers,' by E. W. Grinfield." *Edinburgh Review* xlii (August 1825): 206–23.

Anon. "Charges Delivered by Connop Thirlwall." *Edinburgh Review* 143 (1876): 145–62.

Aristotle. *Aristotle's Politics and Poetics.* Translated by Benjamin Jowett and Thomas Twining. New York: Viking Press, 1957.

Aristotle. *Physics.* Translated by R. P. Hardie and R. K. Gaye. *Works of Aristotle.* Volume 2. Oxford: Clarendon Press, 1930. Available online at www.archive.org.

Arnold, Thomas. *Lectures on Modern History.* New York: D. Appleton and Co., 1845.

Bachelard, Gaston. *Poetics of Space.* Translated by Maria Jolas. New York: Penguin, 1958.

Bacon, Francis. *Novum Organum.* Edited by Joseph Devey. New York: P. F. Collier, 1902. Available online at http://oll.libertyfund.org/titles/bacon-novum-organum

Baker, Robert. *Report of the Leeds Board of Health.* Leeds: Intelligencer Office, 1833.

Bakker, Robert. *Raptor Red.* New York: Bantam Books, 1996.

Ball, Patricia. *Science of Aspects: The Changing Role of Facts in the Work of Coleridge, Ruskin, and Hopkins.* London: Athlone Press, 1971.

Banks, Joseph. *The Endeavour Journal of Sir Joseph Banks*. Edited by J. C. Beaglehole. Sydney: Trustees of Public Library of New South Wales in association with Angus and Robertson, 1962.

Banks, Joseph. *Indian & Pacific Correspondence of Sir Joseph Banks, 1768–1820*, vol. i. Edited by Neil Chambers. London: Pickering & Chatto Ltd., 2008.

Banks, Joseph. "Thoughts on the Manners of the Otaheite." In *Scientific Correspondence of Sir Joseph Banks 1765–1820*, vol. ii, edited by Neil Chambers, appendix iv, 330–37. London: Pickering & Chatto Ltd., 2007.

Barlow, Nora, ed. *Charles Darwin and the Voyage of the Beagle*. London: Pilot Press Ltd., 1945.

Barlow, Nora. "Letter to Nature, Charles Darwin and the Galapagos Islands." *Nature* 136 (September 1935): 391.

Barrow, John. "Dr. Granville's Travels." *Quarterly Review* 39 (1829): 1–41.

Bell, Peter, ed. *Darwin's Biological Work: Some Aspects Reconsidered*. New York: Wiley, 1959.

Bentham, Jeremy. Bentham Library Add. ms 29,809 ff 228–9, Feb. 18, 1819. In Catherine Fuller, "Bentham, Mill, Grote, and *An Analysis of the Influence of Natural Religion on the Temporal Happiness of Mankind*," edited by Kyriakos Demetriou.117–33. *Brill's Companion to George Grote and the Classical Tradition*. Leiden: Brill, 2014.

Bentham, Jeremy. *Comment on the Commentaries and A Fragment on Government*. Edited by J. H. Burns and H. L. A. Hart. Oxford: Oxford University Press, 2009.

Bentham, Jeremy. *Fragment on Ontology. Works of Jeremy Bentham*. Edited by John Bowring. 11 volumes, vol. 8. Edinburgh: William Tait, 1843. Available at online Library of Liberty, www.oll.libertyfund.org.

Bentham, Jeremy. *Handbook of Political Fallacies*. Edited by H. A. Larrabee. New York: Harper, 1962; see also *Works of Jeremy Bentham*. Edited by John Bowring. 11 volumes, vol. 2. Edinburgh: William Tait, 1843. Available at online Library of Liberty, www.oll.liberty-fund.org.

Bentham, Jeremy, *Pannomial Fragments. Works of Jeremy Bentham*. Edited by John Bowring. 11 volumes, vol. 3. Edinburgh: William Tait, 1843. Available at online Library of Liberty, www.oll.libertyfund.org.

Bible: Oxford Annotated, Revised Standard Version. New York: Oxford University Press, 1962.

Brooke, John T. "A Tidal Wave of Disease: The 1832 Leeds Cholera Epidemic." *Bulletin of the Liverpool Medical History Society* 24 (2012–2013): 37–54.

Brougham, Henry. "Speech in the House of Common, on Thursday February 7, 1828, on his motion that an humble Address be presented to his Majesty, praying that he will be graciously pleased to issue a Commission for inquiring into the Defects occasioned by Time or otherwise in the Laws of this Realm, and into the Measures necessary for removing the same." *The Jurist, or the Quarterly Journal of Jurisprudence and Legislation*, vol. ii. London: Baldwin and Craddock, 1828.

Bruner, Jerome. *Culture of Education*. Cambridge, MA: Harvard University Press, 1996.

Bulwer-Lytton, Edward. *England and the English*. 2 volumes. New York: J. & J. Harper, 1833.

Butlin, Martin. "J. M. W. Turner: Art and Content." In *Turner 1775–1851*. Martin Butlin and Andrew Wilton (curators). 9–19. London: Tate Gallery Publications, 1974.

Byron, George Gordon. "Manfred." In David Perkins, *English Romantic Writers*. 812. New York: Harcourt & World, 1967.

Byron, George Gordon, and Thomas Moore. *Works of Lord Byron: with his Letters and Journals, and his Life, by Thomas Moore, esq.* 17 volumes. London: John Murray, 1832–1833.

Byron, John. *The Narrative of the Honourable John Byron (Commodore in a late expedition around the world) Containing an Account of the Great Distresses Suffered by Himself and His Companions on the Coast of Patagonia, from the Year 1740, Till Their Arrival in England. 1746*. London: S. Baker and G. Leigh, and T. Davies, 1768.

Cannon, Susan. *Science in Culture: The Early Victorian Period*. New York: Science History Publications, 1978.

Cannon, W. Faye [Susan]. "The Whewell-Darwin Controversy." *Journal of the Geological Society* cxxxii (1976): 377–84.

Carlyle, Thomas. *Critical and Miscellaneous Essays*. London: James Fraser, 1839.

Carlyle, Thomas. *On Heroes, Hero-Worship and the Heroic in History*. 1st ed., 1840. Oxford: Clarendon Press, 1910.

Carlyle, Thomas. *Sartor Resartus*. 1st ed., 1831. London: Chapman and Hall, 1863.

Chadwick, Edwin. *Report on the Sanitary Conditions of the Labouring Population of Great Britain*. London: Printed by R. Clowes & Sons, for Her Majesty's Stationery Office, 1842. Available online at www.deltaomega.org/documents/ChadwickClassic.pdf.

Chancellor, Gordon, and Randal Keynes. "Darwin's Field Notes on the Galapagos: 'A Little World in Itself.'" Available online at Darwin-online.org.uk/EditorialIntroductions/Chancellor_Keynes_Galapagos.html.

Chapman, Anne. *European Encounters with the Yamana People of Cape Horn, Before and After Darwin*. New York: Cambridge University Press, 2010.

Chaucer, Geoffrey. *The Canterbury Tales*. In *The Norton Anthology of English Literature*, edited by M. Abrams. New York: Norton, 1962.

"Children's Employment Commission, First Report," testimony of Sarah Gooder, aged eight years, #116 in *Parliamentary Papers*, 1842, vols. xv–xvii; available online at https://learn.stleonards.vic.edu.au/yr9hisau/files/2013/02/Sarah-Gooder-Parliamentary-Inquiry-into-Mines-1842.pdf

Clark, John. "Thirlwall." In *Old Friends at Cambridge and Elsewhere*. 77–152. London: Macmillan & Co., 1900.

Clark, John. "Thirlwall, (Newell) Connop (1797–1875)." *Dictionary of National Biography* 56 (1885–1900): 138–41.

Clarke, Martin L. *George Grote: A Biography*. London: The Athlone Press of the University of London, 1962.

Clarke, Martin L. *Greek Studies in England*. Cambridge: Cambridge University Press, 1945.

Clinton, Henry F. *Fasti Hellenici. The Civil and Literary Chronology of Greece, from the Earliest Accounts to the Death of Augustus*. Oxford: Oxford University Press, 1834.

Clive, John. *Macaulay: The Shaping of the Historian*. New York: Vintage, 1973.

Coleman, William. *Georges Cuvier: Zoologist*. Cambridge, MA: Harvard University Press, 1964.

Coleridge, Samuel Taylor. *On the Constitution of Church and State, According to the Idea of Each; with Aids toward a Right judgement of the late Catholic Bill*. London: Hurst, Chance & Co., 1830.

Coleridge, Samuel Taylor. "Rime of the Ancient Mariner." In David Perkins, *English Romantic Writers*. 405–13. New York: Harcourt & World, 1967.

Coleridge. Samuel Taylor. *Statesmen's Manual; or the Bible, The Best guide to Political Skill and foresight: A Lay Sermon addressed to the Higher Classes of Society*. 1816. Reprinted in S. T. Coleridge, *Biographia Literaria and 2 Lay Sermons*. London: G. Bell and Sons, 1894.

Coles, Robert. *The Moral Intelligence of Children*. New York: Random House, 1997.

Collingwood, R. G. *An Autobiography*. London: Oxford University Press, 1939.

Collingwood, R. G. *Idea of History*. London: Oxford University Press, 1946.

Conolly, John. *The Working-Man's Companion: The Physician.1.Cholera*. London: Society for the Diffusion of Useful Knowledge, 1832.

Conybeare, William. "An Examination of Those Phaenomena of Geology, Which Seem to Bear Most Directly on Theoretical Speculations." *Philosophical Magazine and Annals* 8 (1830): 359–62, 401–06; and 9 (1831): 19–23, 111–17, 188–97, 258–70.

Conybeare, William. "On the Hydrographical Basin of the Thames, With a View More Especially to Investigate the Causes Which Have Operated in the Formation of the Valleys of That River, and Its Tributary Streams." *Proceedings of the Geological Society of London* 1, no. 12 (1829): 145–49.

Conybeare, William. "Report on the Progress, Actual State and Ulterior Prospects of Geological Science." *Report of the British Association for the Advancement of Science 1831–2* (1833): 365–414.

Costa, James. "The Darwinian Revelation: Tracing the Origin and Evolution of an Idea." *BioScience* lix (2009): 886–94.

Costa, James. "New Landscapes and New Eyes: The Many Voyages of Charles Darwin." In *(Dis)Entangling Darwin: Cross-Disciplinary Reflections on the Man and His Legacy*, edited by S. da Silva, F. Viera, and J. da Silva. 42–54. Newcastle: Cambridge Scholars Publishing, 2012.

Cowell, Frank. *The Athenaeum: Club and Social Life in London 1824–1974*. Portsmouth, NH: Heinemann Publishers, 1975.

Cuvier, Georges. *Essay on the Theory of the Earth. With Mineralogical Notes and an Account of Cuvier's Geological Discoveries by Professor Jameson. To which are now added Observations on the Geology of North America; illustrated by the description of various organic remains found in that part of the world by Samuel Mitchell*. New York: Kirk & Mercein, 1818.

Cuvier, Georges. "Historical Eloge of the Late Sir Joseph Banks." *Edinburgh New Philosophical Journal* 8 (April 1827): 1–22.

Darrow, Clarence. Closing Argument of Clarence Darrow in the Case of People v. Henry Sweet. In the Recorders Court, Detroit, Michigan. May 11, 1926. Available online at http://law2.umkc.edu/faculty/projects/ftrials/sweet/darrowsummation.html.

Darwin, Charles. *Charles Darwin and the Voyage of the Beagle*. Edited with an introduction by Nora Barlow. London: Pilot Press Ltd., 1945. Available online at http://darwin-online.org.uk/content/frameset?pageseq=1&itemID=F1571&viewtype=text.

Darwin, Charles. *Charles Darwin's Beagle Diary*. Edited by Richard Keynes. Cambridge: Cambridge University Press, 1998.

Darwin, Charles. *Charles Darwin's Notebooks, 1836–1844: Geology, Transmutation of Species, Metaphysical Enquiries*. Cambridge: Cambridge University Press. Available online at http://darwin-online.org.uk/EditorialIntroductions/vanWyhe_notebooks.html.

Darwin, Charles. *Correspondence of Charles Darwin*. Edited by F. Burkhardt and S. Smith. Cambridge: Cambridge University Press, 1985. 24 volumes. Available online in Darwin Correspondence Database, /www.darwinproject.ac.uk/about/publications/correspondence-charles-darwin.

Darwin, Charles. "Darwin's Ornithological Notes," edited by Nora Barlow. *Bulletin of the British Museum (Natural History)*. Historical Series, vol. 2, no. 7, pp. 201–78.

Darwin, Charles. *The Descent of Man, and Selection in Relation to Sex*. London: John Murray, 1871. Available online at http://darwin-online.org.uk/content/frameset?pageseq=1&itemID=F937.1&viewtype=text.

Darwin, Charles. *Journal of Researches into the Geology and Natural History of the Various Countries visited by the H. M. S. Beagle.* London: Henry Colburn, 1839. Available online at http://darwin-online.org.uk/content/frameset?itemID=F10.3&viewtype=text&pageseq=1.

Darwin, Charles. *Life and Letters of Charles Darwin, including an Autobiographical Chapter.* Edited by F. Darwin. London: John Murray, 1887. 3 volumes. Available online at http://darwin-online.org.uk/content/frameset?itemID=F1452.1&viewtype=text&pageseq=1.

Darwin, Charles. *On the Origin of Species by Means of Natural Selection, or the Preservation of Favoured Races in the Struggle for Life.* London: John Murray, 1859. Available online at http://darwin-online.org.uk/content/frameset?itemID=F373&viewtype=text&pageseq=1.

Daubeny, Charles. *Description of Active and Extinct Volcanos,* London: W. Phillips, 1826. Available online at /babel.hathitrust.org/cgi/pt?id=hvd.32044081605321;view=2up;seq=8.

De Botton, Alain. *How Proust Can Change Your Life.* New York: Vintage, 1997.

De Champs, Emmanuelle. "The Place of Jeremy Bentham's Theory of Fictions in 18th Century Linguistic Thought." Available online from the UCL Bentham Project at http://discovery.ucl.ac.uk/647/1/002__1999__EdChamps_1999.pdf.

Demetriou, Kyriakos, ed. *Brill's Companion to George Grote and the Classical Tradition.* Leiden: Brill, 2014.

De Santillana, Giorgio. *Crime of Galileo.* Chicago: University of Chicago Press, 1955.

De Santillana, Giorgio. *Origins of Scientific Thought: From Anaximander to Proclus, 600 b.c. to 300 a.d.* London: Weidenfeld & Nicolson, 1961.

Desmond, Adrian, and James Moore. *Darwin's Sacred Cause: How a Hatred of Slavery Shaped Darwin's Views on Human Evolution.* New York: Houghton Mifflin Harcourt, 2009.

Dickens, Charles. *Bleak House.* Serialized 1852–1853. Book form. London: Bradbury and Evans, 1853. .

Dilke, Charles Wentworth. "Literature of the People." *Athenaeum* 2201 (January 1, 1870): 11–14; see also *Athenaeum*, January–June, 1870. London: John Francis, 1870.

Duhem, Pierre. *Aim and Structure of Physical Theory.* Translated by Philip Wiener. 2nd ed. Princeton: Princeton University Press, 1954. Reprinted, New York: Athenaeum, 1962.

Eastlake, Elizabeth Rigby, Lady. *Mrs. Grote: A Sketch.* London: John Murray, 1880.

Eddington, Arthur. *Nature of the Physical World.* The Gifford Lectures, 1927. New York: The Macmillan Company, 1929.

Eliade, Mircea. *The Forge and the Crucible: The Origins and Structures of Alchemy.* Chicago: University of Chicago Press, 1962.

Emerson, Ralph Waldo. *Representative Men: Seven Lectures.* Boston: Phillips, Sampson, and Co., 1849.

Epstein, James. *Radical Expression: Political Language, Symbol, and Ritual in England, 1790–1850.* Oxford: Oxford University Press, 1994.

Escott, Thomas. *England: Her People, Polity, and Pursuits.* New York: Henry Holt and Co., 1880.

Escott, Thomas. *Gentlemen of the House of Commons.* 2 volumes. London: Hurst and Blackett Ltd., 1902.

Faraday, Michael. *The Chemical History of the Candle.* Reprint of text published by Chautauqua Press in New York ca. 1885–1889. New York: Dover Publications, 2002.

Faraday, Michael. *Experimental Researches in Electricity.* London: Dent, 1914.

Ferguson, Adam. *Essay on the History of Civil Society.* 8th ed. Philadelphia: A. Finley, 1819.

Fitton, William. "Anniversary Address, 1829." *Proceedings* 1 (1834): 133–34.

Fitz-Roy, Robert. *Narrative of the Surveying Voyages of His Majesty's Ships "Adventure" and "Beagle" between the Years 1826 and 1836.* 3 volumes. London: Henry Colburn, 1839. Available online at archive.org/stream/narrativeofsurve02king#page/n9/mode/2up.

Forbes, Duncan. *The Liberal Anglican Idea of History.* Cambridge: Cambridge University Press, 1952.

Forbes, Edward. "On the Connexion between the Distribution of the existing Fauna and Flora of the British Isles, and the Geological changes which have affected their area, especially during the epoch of the Northern Drift." *Memoirs of the Geological Survey,* 336–51. London: Longman, Brown, Green, and Longmans, 1846.

Fox, Charles James. *The Speeches of the Right Honourable Charles James Fox in the House of Commons.* London: Aylott and Jones, 1848.

Fox, Robert. *Caloric Theory of Gases from Lavoisier to Regnault.* Oxford: Clarendon Press, 1971.

France, Anatole. *The Elm-Tree on the Mall.* Translated by B. Drillien. London: John Lane, Bodley Head, 1921.

France, Anatole. *The Garden of Epicurus.* Translated by A. Allinson. London: John Lane, Bodley Head, 1908.

France, Anatole. *Monsieur Bergeret in Paris.* Translated by B. Drillien. London: John Lane, Bodley Head, 1921.

Franklin, Benjamin. "A narrative of the late massacres, in Lancaster County, of a number of Indians, friends of the province, by persons unknown: with some observations of the same." Reprinted in the *London Chronicle,* April 10, 1764, and in the *Gentleman's Magazine,* 1764, vol. 34, pp. 173–78, 23–24, 25–26. Available online at archive.org/stream/narrativeoflatem00fran#page/24/mode/2up.

Fraser, Antonia. *Perilous Question: Reform or Revolution? Britain on the Brink, 1832.* London: Weidenfeld & Nicolson, 2013.

Froude, Hurrell. *The Remains of the Late Reverend Richard Hurrell Froude.* 2 volumes. London: J. G. and F. Rivington, 1838.

Fuller, Catherine. "Bentham, Mill, Grote, and *An Analysis of the Influence of Natural Religion on the Temporal Happiness of Mankind.*" In *Brill's Companion to George Grote and the Classical Tradition,* edited by Kyriakos Demetriou. 117–33. Leiden: Brill, 2014. Available online at discovery.ucl.ac.uk.

Galileo. *Discoveries and Opinions of Galileo.* Translated with an introduction by Stillman Drake. New York: Anchor Books, 1957.

Garland, Martha M. *Cambridge Before Darwin: The Ideal of a Liberal Education.* Cambridge: Cambridge University Press, 1981.

Gellner, Ernest. *Legitimation of Belief.* Cambridge: Cambridge University Press, 1974.

Gillispie, Charles. *The Edge of Objectivity: An Essay in the History of Scientific Ideas.* Princeton, NJ: Princeton University Press, 1960.

Godwin, William. *Things as They Are; or the Adventures of Caleb Williams.* 3 volumes. London: B. Crosby, 1794.

Gould, Stephen Jay. *The Panda's Thumb.* New York: Norton, 1980.

Gregory, Kenneth. *The Second Cuckoo: A Further Selection of Witty, Amusing and Memorable Letters to the Times.* London: George Allen and Unwin, 1983.

Grote, George [Philip Beauchamp]. *An Analysis of the Influence of Natural Religion on the Temporal Happiness of Mankind.* London: R. Carlile, 1822.

Grote, George. "The Essentials of Parliamentary Reform." 1st ed. London: Baldwin and Craddock, 1831. Reprinted in *Minor Works of George Grote,* edited by Alexander

Bain. 1–55. London: John Murray, 1873. Available online at archive.org/details/minorworksofgeor00grotuoft.

Grote, George. "Fasti Hellenici." *Westminster Review* 5 (1826): 269–331; see also *The Minor Works of George Grote*, edited by Alexander Bain. 13–18. London: John Murray, 1873.

Grote, George. "Grecian Legends and Early History." *Westminster Review* 39 (1843): 285–328. Reprinted in *Minor Works of George Grote*, edited by Alexander Bain. 73–134. London: John Murray, 1873. Available online at archive.org/details/minorworksofgeor00grotuoft.

Grote, George. *A History of Greece; from the Earliest Period to the Close of the Generation Contemporary with Alexander the Great*. 12 volumes. London: John Murray, 1869.

Grote, George. *The Minor Works of George Grote*. Edited by Alexander Bain. London: John Murray, 1873. Available online at archive.org/details/minorworksofgeor00grotuoft.

Grote, Harriet. *Memoir of the Life of Ary Scheffer*. London: John Murray, 1860.

Grote, Harriet. *The Personal Life of George Grote*. London: John Murray, 1873.

Hall, Captain Basil. "Letter from Captain Basil Hall, R. N. to Captain Kater, Communicating the Details of Experiments." Read April 24, 1823. *Philosophical Transactions of the Royal Society* 113 (1823): 211–85. Available online at archive.org/stream/jstor-107651/107651_djvu.txt.

Hammond, John, and Barbara Hammond. *Lord Shaftesbury*. New York: Harcourt, Brace and Co., 1923.

[Hare, J. C., and A. W. Hare.] *Guesses at Truth by Two Brothers*, 3rd ed. 1st ed., 1827. London: Macmillan and Co., 1867.

Havelock, Eric. *Preface to Plato*. Cambridge, MA: Harvard University Press, 1963.

Hay, Douglas, ed. *Albion's Fatal Tree: Crime and Society in Eighteenth-Century England*. New York: Pantheon Books, 1975.

Herbert, Sandra. "Charles Darwin as a Prospective Geological Author." *British Journal for the History of Science* 24 (1991): 159–92.

Herschel, John. "Letter to Lyell, Feb. 20, 1836." In *Ninth Bridgewater Treatise: A Fragment*, by Charles Babbage, appendix, note 1. 2nd ed. London: John Murray, 1838.

Herschel, John. *A Preliminary Discourse on the Study of Natural Philosophy*. Dionysius Lardner's Cabinet Cyclopaedia, vol. i. London: Longman, Rees, Orme, Brown & Green and John Taylor, 1831.

Hilgers, Lauren. "Hong Kong's Umbrella Revolution Isn't Over Yet," *New York Times Magazine*. February 22, 2015. Available online at www.nytimes.com/2015/02/22/magazine/hong-kongs-umbrella-revolution-isnt-over-yet.html?ref=magazine.

Himmelfarb, Gertrude. *The Spirit of the Age: Victorian Essays*. New Haven, CT: Yale University Press, 2007.

Hindley, Charles. "An Address, delivered at the Establishment of the Mechanics' Institutions, Ashton-under-Lyme June 22, 1825." *Edinburgh Review* xlii (August 1825): 499–504. Available online at babel.hathitrust.org/cgi/pt?id=inu.30000093204380;view=2up;seq=8.

Holmes, Richard. *The Age of Wonder*. New York: Pantheon Books, 2008.

Howarth, Osbert. *The British Association for the Advancement of Science: A Retrospect 1831–1931*, London: The Association, 1931.

Hooykaas, Reijer. *Natural Law and Divine Miracle: The Principle of Uniformity in Geology, Biology, and Theology*. Leiden: Brill, 1963.

Huxley, Thomas H. "On the Reception of the Origin of Species." In *Life and Letters of Charles Darwin*, edited by F. Darwin. 3 volumes. Volume 2. 179–205. London: John Murray, 1887. Available online at http://darwin-online.org.uk/content/frameset?itemID=F1452.2&viewtype=text&pageseq=1.

Irvine, William. *Apes, Angels, & Victorians: A Joint Biography of Darwin & Huxley*. London: Readers Union, 1956.

Jackson, James B. *The Necessity for Ruins and Other Topics*. Amherst: University of Massachusetts Press, 1980.

Jefferson, Thomas. Correspondence. "Letter to William Ludlow, September 6, 1824." Available online through a project at the University of Groningen, The Netherlands at http://www. let.rug.nl/usa/presidents/thomas-jefferson/letters-of-thomas-jefferson/jefl279.php.

Jefferson, Thomas. *Notes on the State of Virginia*. Edited with an Introduction and Notes by William Peden. 1st ed. 1785. Chapel Hill: University of North Carolina Press, 1955.

Jones, Jonathan. "Turner and Constable Exhibitions Revive Britain's Greatest Art Rivalry." *Guardian*, August 24, 2014. Available online at www.goodreads.com/author_blog_posts/ 6876897-turner-and-constable-exhibitions-revive-britain-s-greatest-art-rivalry.

Kargon, Robert. *Science in Victorian Manchester: Enterprise and Expertise*. Baltimore: Johns Hopkins University Press, 1977.

Kay, James. *Moral and Physical Condition of the Working Class Employed in the Cotton Manufacture in Manchester*. London: James Ridgway, 1832. Available online at archive. org/details/moralphysicalcon00kaysuoft.

Kettel, Thomas. "Law Reform in England." *The United States Magazine and Democratic Review* xxviii (1851): 32–48.

Kierstead, James. "Grote's Athens: The Character of Democracy." In *Brill's Companion to George Grote and the Classical Tradition*, edited by Kyriakos Demetriou. 161–210. Leiden: Brill, 2014.

Kormondy, Edward. *Concepts of Ecology*. Englewood Cliffs, NJ: Prentice-Hall, 1969.

Kummel, Werner. *New Testament: The History of the Investigation of its Problems*. New York: Abingdon Pres, 1972.

Kurlansky, Mark, ed. *Choice Cuts: A Savory Selection of Food Writing from Around the World and Throughout History*. New York: Penguin, 2002.

Ladd, Henry. *Victorian Morality of Art*. New York: R. Long and R. R. Smith, Inc., 1932.

Lakatos, Imre. *Proofs and Refutations*. Cambridge: Cambridge University Press, 1976.

Lamb, Charles. "The Superannuated Man." In *The Last Essays of Elia*, 1st ed. London: Edward Moxon, 1833. In *The Complete Works and Letters of Charles Lamb*, introduction by Saxe Commins. New York: Modern Library, 1935.

Levine, Joseph. *Dr. Woodward's Shield: History, Science, and Satire in Augustan England*. Ithaca, NY: Cornell University Press, 1991.

Linebaugh, Peter. "The Tyburn Riot against the Surgeons." In *Albion's Fatal Tree: Crime and Society in Eighteenth-Century England*, edited by Douglas Hay. 65–117. New York: Pantheon Books, 1975.

Lyell, Charles. *Life, Letters and Journals of Sir Charles Lyell*. Edited by Katharine Lyell. 2 volumes. London: John Murray, 1881.

Lyell, Charles. *Principles of Geology: Being An Attempt to Explain the Former Changes of the Earth's Surface by Reference to Causes Now in Operation*. London: John Murray, 1830.

Macaulay, Thomas Babbington. *History of England from the Accession of James the Second*. 1st ed. 1848. 5 volumes. London: Longmans, Green, and Co., 1886.

Macaulay, Thomas Babbington. "Parliamentary Address of March 2, 1831." In *Works of Lord Macaulay* , vol. x, 425–26. Albany edition. 12 volumes. London: Longmans, Green and Co., 1898.

Macaulay, Thomas Babbington. "Speech in Support of the Anatomy Act," delivered February 27, 1832. In *Miscellaneous Writings and Speeches of Lord Macaulay*, vol. iv. London: Longmans, Green Co., 1889. Available online at www.gutenberg.org/files/2170/2170-h/2170-h.htm.

MacCormmach, Russell. *Night Thoughts of a Classical Physicist*. Cambridge, MA: Harvard University Press, 1982.

Malamud, Bernard. *The Magic Barrel*. Philadelphia: The Jewish Public Society of America, 1953.

Marks, Richard. *Three Men of the Beagle*. New York: Knopf, 1991.

Marshall, Fiona. "History of the Philological Society: The Early Years." 2006. Available online at file:///C:/Users/Louis/Downloads/Early_PhilSoc_history%20(3).pdf.

Martineau, Harriet. *How to Observe. Morals and Manners*. London: Charles Knight, 1838.

Measureworth.com. Available at https://www.measuringworth.com/ukcompare/.

Menand, Louis. *Metaphysical Club*. New York: FSG, 2001.

Mill, John Stuart. *Autobiography*. London: Longmans, Green, Reader, and Dyer, 1873; see also *The Collected Works of John Stuart Mill*, edited by J. M. Robson. 33 volumes. Vol. i, 5–290. Available online at http://oll.libertyfund.org/titles/mill-the-collected-works-of-john-stuart-mill-volume-i-autobiography-and-literary-essays.

Mill, John Stuart. *Autobiography, Early Draft*. In *The Collected Works of John Stuart Mill*, edited by J. M. Robson. 33 volumes. Vol. i, 4–248. Available online at http://oll.libertyfund.org/titles/mill-the-collected-works-of-john-stuart-mill-volume-i-autobiography-and-literary-essays.

Mill, John Stuart. *Collected Works of John Stuart Mill*. Edited by J. M. Robson. 33 volumes. Toronto: University of Toronto Press, 1963–1991. Available at http://oll.libertyfund.org/titles/mill-collected-works-of-john-stuart-mill-in-33-vols.

Mill, John Stuart. "Grote's *History of Greece*." *Edinburgh Review* 84 (1846): 343–77. Available online at http://oll.libertyfund.org/titles/mill-the-collected-works-of-john-stuart-mill-volume-xi-essays-on-philosophy-and-the-classics.

Mill, John Stuart. "The Literary Remains of Samuel Taylor Coleridge." *London and Westminster Review* xxxiii (March 1840): 257–302; see also *Collected Works of John Stuart Mill*, edited by J. M. Robson. 33 volumes. Vol. x, 119–64. Available online at http://oll.libertyfund.org/titles/mill-the-collected-works-of-john-stuart-mill-volume-x-essays-on-ethics-religion-and-society

Mill, John Stuart (anon.). "Notes on the Newspapers." *Monthly Repository* N.S.8 (August 1834): 589–600, in *Collected Works of John Stuart Mill*, vol. vi. Available online at http://files.libertyfund.org/files/245/0223.06_Bk.pdf.

Mill, John Stuart. "Spirit of the Age." *Political Examiner*. There are 7 instalments: January 9, 1831, 20–21; January 23, 1831, 50–52; February 6, 1831, 82–84; March 13, 1831, 162–63; April 3, 1831, 210–11; May 15, 1831, 307, and May 29, 1831, 339–41 in *Collected Works of John Stuart Mill*, vol. xxii. These are all available online at http://oll.libertyfund.org/titles/mill-the-collected-works-of-john-stuart-mill-volume-xxii-newspaper-writings-part-i.

Mill, John Stuart. "The Works of Jeremy Bentham." *London and Westminster Review* 7 and 29 (August 1838): 467–506; see also *Collected Works of John Stuart Mill*, edited by J. M. Robson. 33 volumes. Vol. x, 77–115. Available online at http://oll.libertyfund.org/titles/mill-the-collected-works-of-john-stuart-mill-volume-x-essays-on-ethics-religion-and-society.

Milman, Henry. "Grote's *History of Greece*." *Quarterly Review* 78 (1846): 113–44.

Mitford, William. *History of Greece*. 5 volumes. London: T. Caddell and W. Davies, 1784–1810.

Momigliano, Arnaldo. *Essays in Ancient and Modern Historiography*. Middletown, CT: Wesleyan University Press, 1977.

Momigliano, Arnaldo. *Studies in Historiography*. London: Weidenfeld and Nicolson, 1966.

Morgan, John. "Bishop Thirlwall." In *Four Biographical Sketches*. 63–111. London: Elliot Stock, 1892.

Murray, Gilbert. *Aeschylus*. London: Clarendon Press, 1940.

Newton, Isaac. "De Gravitatione" (ca. 1666). In *Unpublished Scientific Papers of Isaac Newton*, edited and translated by Rupert Hall and Marie Boas Hall. 89–156. Cambridge: Cambridge University Press, 1962.

Newton, Isaac. *Mathematical Principles of Natural Philosophy and His System of the World*. 2 volumes. Revised. Berkeley: University of California Press, 1966.

Niebuhr, Barthold Georg. *History of Rome*. Translated by J. C. Hare and C. Thirlwall. London: Taylor & Walton, 1827.

Neibuhr, Barthold Georg. *The Life and Letters of Barthold Georg Niebuhr, and Selections from his Minor Writings*. Edited by Susanna Winkworth. 3 volumes. London: Chapman and Hall, 1852.

Paradis, James, and Thomas Postlewait, eds. *Victorian Science and Victorian Values: Literary Perspectives*. New York: New York Academy of Sciences, 1981.

The Penny Cyclopaedia of the Society for the Diffusion of Useful Knowledge. 27 volumes. London: Charles Knight and Co., 1833–1838.

Pfeiffer, John. *Emergence of Man*. New York: Harper & Row, 1972.

Phillips, John. *Illustrations of the Geology of Yorkshire; or A Description of the Strata and Organic Remains of the Yorkshire Coast*. York: Thomas Wilson and Sons, 1829.

Phillips, John. *Illustrations of the Geology of Yorkshire. Part II, Mountain Limestone District*. London: John Murray, 1836.

Place, Francis. *London Radicalism from 1830–1843: A Selection from the Papers of Francis Place*. Edited by D. J. Rowe. Vol. v, 218–36. London: London Records Society Publications, 1970. Available online at http://www.british-history.ac.uk/london-record-soc/vol5.

Plato. *Phaedrus*. Translated with an introduction and commentary by R. Hackforth. Cambridge: Cambridge University Press, 1952.

Plato. *Timaeus and Critias*. Translated with an introduction and an appendix on Atlantis by Desmond Lee. Harmondsworth: Penguin, 1971.

Playfair, John. *Illustrations of the Huttonian Theory of the Earth*. Edinburgh: William Creech, 1802.

Popper, Karl. *The Open Society and its Enemies*, vol. i. London: Routledge and Kegan Paul, 1945.

Popper, Karl. "On the Sources of Knowledge and Ignorance," In *Conjectures and Refutations*. 2–30. London: Routledge & Kegan Paul, 1963.

Rankin, Ian. *The Falls*. London: Orion Publishing Group, 2000.

Report from the Committee of Secrecy on the Bank of England Charter. London: James & Luke G. Hansard & Sons, 1832.

Richardson, Sarah. "A Regular Politician in Breeches: The Life and Work of Harriet Lewin Grote." In *Brill's Companion to George Grote*, edited by Kyriakos Demetriou. Leiden: Brill, 2014.

Robertson, George C. "George Grote." In *Dictionary of National Biography* 23 (1885–1900): 284–93. Available online at en.wikisource.org/wiki/Grote,_George_(DNB00).

Rosenblatt, Louis. "Fossils and Myths." Dissertation. Johns Hopkins University, 1983.

Rosenblatt, Louis. *Rethinking the Way We Teach Science: The Interplay of Content, Pedagogy, and the Nature of Science*. New York: Routledge, 2011.

Rosenhek, Jackie. "Invasion of the Body Snatchers." *Doctor's Review* (May 2005). Available online at http://www.doctorsreview.com/history/may05_history/.

Rudwick, Martin. *The Great Devonian Controversy*. Chicago: University of Chicago Press, 1985.

Rudwick, Martin. "Introduction." In *Principles of Geology, being an Attempt to Explain the Former Changes of the Earth's Surface, by reference to Causes now in Operation*, by Charles Lyell. Reprint of 1st ed. i–lviii. Chicago: University of Chicago Press, 1990.

Rudwick, Martin. *The Meaning of Fossils: Episodes in the History of Palaeontology*. London: MacDonald, 1972.

Ruskin, John. *Modern Painters*. 5 volumes. New York: Wiley, 1862.

Sandburg, Carl. *Complete Poems of Carl Sandburg*. Introduction by Archibald MacLeish. New York: Harcourt, 1970.

Schleiermacher, Friedrich. *A Critical Essay on the Gospel of St. Luke*. With an introduction by the translator, Connop Thirlwall. London: John Taylor, 1825.

Schweber, S. S. "Scientists as Intellectuals: The Early Victorians." In *Victorian Science and Victorian Values: Literary Perspectives*, edited by James Paradis and Thomas Postlewait. 1–37. New York: New York Academy of Science, 1981.

Sedgwick, Adam. "Anniversary Address, 1830." *Proceedings of the Geological Society of London* 1 (1834): 187–213.

Sedgwick, Adam. "Anniversary Address, 1831." *Proceedings of the Geological Society of London* 1 (1834): 281–317.

Sedgwick, Adam. *A Discourse on Studies of the University*. Reprint of 1st ed. (1833), with an introduction by Eric Ashby and Mary Anderson. New York: Humanities Press and Leicester University Press, 1969.

Sedgwick, Adam. *A Discourse on Studies of the University*. 5th ed. London: John W. Parker, 1850.

Sedgwick, Adam. *Life and Letters of the Reverend Adam Sedgwick*. 2 volumes. Edited by John W. Clark and Thomas M. Hughes. Cambridge: Cambridge University Press, 1890.

Sedgwick, Adam, and Roderick Murchison. "A Sketch of the Structure of the Eastern Alps; with Sections through the Newer Formations on the Northern Flanks of the Chain, and through the Tertiary Deposits of Styria, etc." *Transactions of the Geological Society of London*, New Series iii (1835): 301–420.

Sherwood, Mary. "The Penny Tract." In *The Works of Mrs. Sherwood*. 15 volumes. Vol. xiii, 350–62. New York: Harper & Brothers, 1837.

Sleeman, Lieutenant Colonel William. *Rambles and Recollections of an Indian Official*. 2 volumes. London: Hatchard and Son, 1844.

Smith, Adam. *Inquiry into the Nature and Causes of the Wealth of Nations*. 1st ed., 1776. 2 volumes. Edited with an introduction, notes, marginal summary and enlarged index by Edwin Cannan. London: Methuen, 1904

Smith, Sydney. "Bentham on Fallacies." Originally published in *Edinburgh Review*, 1825. In *Works of the Rev. Sydney Smith*, vol. iii, 209–19. Philadelphia: Carey and Hart, 1845.

Snell, Bruno. *Discovery of the Mind: The Greek Origins of European Thought*. 2nd ed. Translated by T. G. Rosenmeyer. Oxford: Basil Blackwell, 1948.

Snow, John. *On the Mode of Communication of Cholera*. London: John Churchill, 1855.

Snow, William Parker. *A Two Year's Cruise off Tierra del Fuego, the Falkland Islands, Patagonia, and in the River Plate: A Narrative of Life in the Southern Seas*. London: Longman, Brown, Green, Longmans, & Roberts, 1857.

Snyder, Laura. "'The Whole Box of Tools': William Whewell and the Logic of Induction." In *British Logic in the Nineteenth Century*, edited by Dov Gabbay and John Woods. 163–228.

Amsterdam: Elsevier, 2008. Available online at http://uwch-4.humanities.washington. edu/Texts/JOSH-H/Philosophy%20Guides,%20Analysis'%20and%20Resources%20 (ver.2)/British%20Logic%20in%20the%20Nineteenth%20Century,%20Volume%204%20 (Handbook%20of%20the%20History%20of%20Logic).pdf.

Spencer, Herbert. *Essays: Scientific, Political, & Speculative.* 3 volumes. Vol. i, 1–7. London: Williams and Norgate, 1891. Available online at http://oll.libertyfund.org/titles/ spencer-essays-scientific-political-and-speculative-vol-1--5.

Stanley, Arthur. *Sermons on Special Occasions.* London: John Murray, 1882.

Strachey, Lytton. *Eminent Victorians.* New York: Harcourt Brace, 1918.

Storella, Elaine. *"O, what a World of Profit and Delight": The Society for the Diffusion of Useful Knowledge.* Waltham, MA: Brandeis University Press, 1969.

Subercaseux, Benjamin. *Jemmy Button.* New York: The Macmillan Company, 1954.

Sulloway, Frank. "Darwin and His Finches: The Evolution of a Legend." *Journal of the History of Biology* 15, no. 1 (1982): 1–53.

Sulloway, Frank. "Darwin's Conversion: The Beagle Voyage and Its Aftermath." *Journal of the History of Biology* 15, no. 3 (Fall 1982): 325–96.

Thirlwall, Connop. *History of Greece.* The Cabinet of History conducted by the Rev. Dionysius Lardner. London: Longman, Rees, Orme, Brown, Green and Longman and John Taylor, 1835.

Thirlwall, Connop. "Kruse's *Hellas.*" *Philological Museum* 1 (1832): 305–58.

Thirlwall, Connop. *"A Letter to Thomas Turton on the Admission of Dissenters to Academical Degrees,"* 1–45. Cambridge: J. & J. J. Deighton, 1834.

Thirlwall, Connop. *Letters Literary and Theological of Connop Thirlwall.* Edited by J. J. Stewart Perowne and Louis Stokes. London: Richard Bentley & Son, 1881.

Thirlwall, Connop. *Letters to a Friend.* Edited by the Very Rev. Arthur Penrhyn Stanley, D. D. London: Richard Bentley & Son, 1881.

Thirlwall, John. *Connop Thirlwall: Historian and Theologian.* London: Society for Promoting Christian Knowledge, 1936.

Thomas, Dylan. *Collected Stories.* New York: New Directions Books, 1984.

Thomas, Dylan. *The Doctor and the Devils.* London: J. M. Dent, 1953.

Thomson, William. *Making of the English Working Class.* New York: Pantheon Books, 1964.

Thoreau, Henry David. *Walden.* New York: Thomas Y. Crowell, & Co., 1910.

Thornbury, Walter. *Old and New London.* London: Cassell, Peter, & Galpin, 1873.

Timbs, John. *Clubs and Club Life in London: With Anecdotes of Its Famous Coffee Houses, Hostelries and Taverns from the Seventeenth Century to the Present Time.* London: John Camden Hotten, 1872.

Timbs, John. *Curiosities of London: Exhibiting the Most Rare and Remarkable Objects of Interest in the Metropolis; with Nearly Fifty Years Personal Recollections.* London: David Bogue, 1855.

Todhunter, Isaac. *William Whewell, D. D., Master of Trinity College. An Account of his Writings and Selections from his Literary and Scientific Correspondence.* 2 volumes. London: Macmillan and Co., 1876.

Trevelyan, George Macaulay. *Illustrated English Social History,* vol. iii, *The Eighteenth Century.* London: Longmans, Green and Co., 1944.

Tyndall, John. *Faraday as a Discoverer.* London: Longmans, Green and Co., 1894.

Updike, John. "1940's." In *Five Boyhoods,* edited by Martin Levin. 176–98. New York: Doubleday & Company, Inc., 1962.

Vale, Edmund. *Curiosities of Town and Countryside*. London: B. T. Batsford Ltd., 1941.

Van der Valk, Arnold. "Origins and Development of Ecology." In *Philosophy of Ecology*, edited by Kevin de laPlante, Bryson Brown, and Kent Peacock. 25–48. Amsterdam: Elsevier, 2011.

A Vision of Britain Through Time. A website on census data. Available at http://www.vision-ofbritain.org.uk/.

Wallas, Graham. *Life of Francis Place*. London: Longmans, Green, and Co., 1898.

Ward, Thomas H. *History of the Athenaeum: 1824–1925*. London: Printed for the Club, 1926.

Weiner, Jonathan. *Beak of the Finch: A Story of Evolution in Our Time*. New York: Random House, 1991.

Whewell, William. "Lyell—Principles of Geology." *British Critic, Quarterly Theological Review, and Ecclesiastical Record* 9 (January 1831): 180–206.

Whewell, William. "Anniversary Address of 1838." *Proceedings of the Geology Society* ii, no. 55 (1838): 624–49.

Whewell, William. *William Whewell, D. D., Master of Trinity College. An Account of his Writings and Selections from his Literary and Scientific Correspondence*. Edited by Isaac Todhunter. London: Macmillan and Co., 1876.

White, Terence H. *Age of Scandal: An Excursion through a Minor Period*. London: Jonathan Cape, 1950.

Williams, L. Pierce. *Michael Faraday: A Biography*. New York: Basic Books, 1965.

Wilson, John. *Imperial Gazetteer of England and Wales*. 6 volumes. London: A. Fullarton & Co., 1870–1872.

Winkworth, Susanna. *Life and Letters of Barthold Georg Niebuhr, and Selections from his Minor Writings*. 3 volumes. London: Chapman and Hall, 1852.

Wyhe, John van. "Introduction to Charles Darwin's *Journal*, 1809–1881." Available online at http://darwin-online.org.uk/EditorialIntroductions/vanWyhe_JournalDAR158.html.

Wyhe, John van. "Mind the Gap." Available online at http://darwin-online.org.uk/content/frameset?viewtype=text&itemID=A544&pageseq=1.

Wyhe, John van. "Where Do Darwin's Finches Come From?" *Evolutionary Review: Art, Science, Culture* iii, no. 1 (2002): 185–95.

Wyhe, John van, and Peter Kjaegaard. "Going the Whole Orang: Darwin, Wallace, and the Natural History of Orangutans." *Studies in History and Philosophy of Science. Part C Studies in History and Philosophy of Biological and Biomedical Sciences* li (June 2015): 53–63.

INDEX

Abraham, 165, 185
Abrams, M. H., 134
Aldeburgh, Rye, 109–14, 150
Anatomy Act, 7–11
Anaximander, 13, 63
Aristotle, 39, 55, 59, 63, 65–67, 101, 126,
    129, 156, 170, 182–83
Ashley, Lord Anthony, 77–78
Athenaeum, 32, 55–56, 79, 103

Bachelard, Gaston, 181, 184
Bacon, Francis, 80, 176
Baker, Robert, 6–8, 79
Bank of England, 27–28, 56
Banks, Joseph, 140–44, 155–57, 165, 168
Bentham, Jeremy
    fictions, theory of, 68, 121–24, 128
    influence, 20–21, 23, 28, 36, 39, 53, 55, 60,
        68, 108, 120, 126–30
    life, 38, 39, 53, 56, 121
Bergeret. *See* France, Anatole
Bible & Biblical Criticism
    accommodation, principle of, 47,
        87–88, 89–90
    bibliolatry, 88, 90
    inspiration (*see* Schleiermacher; Thirlwall)
British Association for the Advancement of
    Science, 76, 82–83, 90, 91, 135
Brougham, Lord Henry, 126–27
Bulwer-Lytton, Edward (Lord Bulwer
    Lytton), 2, 19, 33, 84
Burdett, Francis, 22
Button, Jemmy, 145–46, 154, 156,
    167, 172–73
Byron, Lord John, 155–56, 164, 175

Cannon, Susan (Cannon, Walter F.),
    95, 117n25
Carlyle, Thomas
    On Heroes, 48, 64, 73, 76, 106, 183
    Sartor Resartus, 72–73
    "Signs of the Times," 179–82, 183
Chadwick, Edwin, 7–8
Chaucer, Geoffrey, 147–48
cholera, 6–8, 20, 79, 149, 178
Clarke, Martin L., 39, 53
Coleridge, Samuel Taylor, 37, 39, 55, 71, 76,
    88, 116, 161, 182
    ideas, theory of, 130–33
    as a lamp, 134–35, 136, 137
Collingwood, R. G., 104, 161–62
Conditions of Working Class, 7, 10–11, 20,
    77–78, 80–81, 180–81
Conolly, John, 6, 7, 9
Constable, John, 11–12, 95
Conybeare, William, 91, 92, 110–13
Cook, Captain James, 140–44, 157, 167
Costa, James, 163
Cuvier, Georges
    Cuvierian approach, 92–95, 100, 101, 105,
        116, 182
    geology, 85–87, 88, 91, 103, 108, 114–15,
        133, 148, 149, 166

Darwin, Charles
    evolution, theory of, 147–49
    Fuegians, 144, 146, 153–57, 163–68,
        173, 184
    Galapagos and Gould, 159–63, 166,
        171, 172
    geology, 90, 101, 102, 106, 108, 149–52